Recycling Class

Urban and Industrial Environments

Series editor: Robert Gottlieb, Henry R. Luce Professor of Urban and Environmental Policy, Occidental College

Recycling Class

The Contradictions of Inclusion in Urban Sustainability

Manisha Anantharaman

The MIT Press
Cambridge, Massachusetts
London, England

The MIT Press
Massachusetts Institute of Technology
77 Massachusetts Avenue, Cambridge, MA 02139
mitpress.mit.edu

The MIT Press would like to thank the anonymous peer reviewers who provided comments on drafts of this book. The generous work of academic experts is essential for establishing the authority and quality of our publications. We acknowledge with gratitude the contributions of these otherwise uncredited readers.

This book was set in Stone Serif and Stone Sans by Westchester Publishing Services.

Library of Congress Cataloging-in-Publication Data

Names: Anantharaman, Manisha, author.
Title: Recycling class : the contradictions of inclusion in urban sustainability / Manisha Anantharaman.
Description: Cambridge, Massachusetts : The MIT Press, [2023] | Series: Urban and industrial environments | Includes bibliographical references and index.
Identifiers: LCCN 2022061580 (print) | LCCN 2022061581 (ebook) | ISBN 9780262546973 (paperback) | ISBN 9780262376983 (epub) | ISBN 9780262376976 (pdf)
Subjects: LCSH: Recycling industry—India—Bangalore. | Recycling (Waste, etc.)—India—Bangalore. | Environmentalism—India—Bangalore. | Community development—India—Bangalore. | Sustainable development—India—Bangalore. | Bengaluru (India)—Economic conditions.
Classification: LCC HD9975.I43 B363 2023 (print) | LCC HD9975.I43 (ebook) | DDC 363.72/8209548—dc23/eng/20230125
LC record available at https://lccn.loc.gov/2022061580
LC ebook record available at https://lccn.loc.gov/2022061581

150507879

For all my patient canine companions, past, present, and future

Contents

Acknowledgments

My mother is a journalist and writer. It is to her I owe my biggest debt. She modeled tenacity and determination, writing two books and hundreds of newspaper articles all while juggling a full-time job, two children who were always sick, family demands, oppressive patriarchy, and a debilitating illness. While I often resented being dragged into her writing projects as a reluctant fact-checker and editor, I now look back on those times as when I first learned to write for the reader. Amma, thank you. I know you might never read this book, but it would not have been written without you.

This book emerged out of a dissertation project that morphed into a longer-term commitment. There were many people, several featured in these pages, whose time, energy, and insight not only helped me develop my research, but also sustained me with their friendship. You are all unnamed but have my deepest gratitude. To the staff and leaders of the civil society organizations and labor groups I engaged in this research, I owe many thanks, for trusting me with your data and goals, for your willingness to have difficult conversations, and for your openness to learning together. In my analysis of the politics of waste in Bengaluru, I have tried to remain accountable to our collaboration and relationship over the years.

At the early stages of this work, I was lucky to have conversations with Vinay Gidwani, Solomon Benjamin, Leo Saldanha, Raka Ray, Isha Ray, and Carol Upadhya, who pushed me to deepen and expand my inquiry. They introduced me to the scholarship on economic and urban geography, cultural sociology, southern urbanism, and labor studies in the kindest way possible. Thanks also to Alastair Iles and Dara O'Rourke, who had faith in my capacities to learn how to be a social scientist and agreed to chair my dissertation, even though I was a total novice in the environmental social sciences. My apologies to the soil microbes I abandoned in the lab in Building 80 of Lawrence Berkeley National Lab. I meant to come back for you, but I got carried away by all the consumption and waste.

I had almost given up on this book project. Three people were critical to helping me persist. Kate O'Neill, who has been a reliable mentor and friend from the time I took her course my first semester of graduate school, reviewed my proposal and encouraged me to keep going. My partner Naïm sent out follow-up emails to editors on my behalf when I was too dejected to try again (and it worked!). Weekly writing sessions with Jennifer Tucker, Kristen Nelson, and Lani Tsinnajinnie were sacred containers for my writing. Many thanks also to Audra J. Wolfe, who helped me learn the ins and outs of book publishing and provided excellent editorial support along the way.

A wonderful community of colleagues and friends reviewed drafts of the chapters in this book and provided invaluable feedback. First and foremost, Jennifer Tucker and John Stehlin—my rockstar writing group (I still can't believe you let me be in a writing group with you). Y'all are brilliant, generous, and insightful. It has meant so much to know I can count on your consistent feedback, commitment, reliably quick turnarounds, and cheerleading. Some of the best sentences and ideas in this book are a product of the many Zoom conversations that we have had over the past six years. Conversations with Freyja Knapp, Margot Higgins, and Sibyl Diver have helped frame and enrich this work in profound ways. To my younger brother Harsha, whose deep knowledge, insight, analysis, and humor I have had the unbelievable luck to benefit from since age four, thank you for reading the final version of all these chapters and, more importantly, for all the conversations about waste, labor, space, and cities over the years. Here's to forever confusing all of academia with our H. Anantharaman and M. Anantharaman signatures!

My deep gratitude also to Hemangini Gupta, Smriti Srinivas, and other members of the Bengaluru Research Network who created space for me to workshop book chapters. Aman Luthra has also been a consistent companion in my intellectual forays into economic geography. Early versions of these book chapters were presented and workshopped at the University of California Berkeley's Global Metropolitan Studies cluster, University of Geneva, University of Manchester, University of Leeds, University of California Santa Barbara, University of Utah, the Relational Poverty Network's conference at the University of Washington, and at Stanford University. Special thanks to Marlyne Sahakian and the Alba Viotto Professorship at the Institute of Sociological Research at the University of Geneva, Tullia Jack and the Department of Service Studies at Lund University, and Alison Browne and the Sustainable Consumption Institute at the University of Manchester for intellectually stimulating and soul-nourishing research stays at their respective institutions. I am further indebted to Beth Clevenger, Anthony Zannino, and Susan Campbell at the MIT Press whose guidance and attention helped see this book to the finish line. I also want to extend a special thank-you to the four anonymous reviewers, whose incisive feedback and generous engagement improved the work immeasurably.

At Saint Mary's College of California, Michael J. Viola and Suzanne Schmidt have helped me learn and grow in my understanding of critical theories and racial capitalism. Monica Fitzgerald, Alicia Rusoja, and Tamara Spencer offered support and encouragement when I needed it the most. Many thanks also to the cohorts of students in my Environmental Justice, Sustainability, and Global Studies classes at Saint Mary's, whose curiosity, honesty, and vulnerability have touched me in profound ways and helped me become a better scholar. Several student research assistants over the years have helped transcribe and code interviews, compile newspaper articles, and check references. Provost Research Grants supported some of the follow-up fieldwork for this book.

An early version of chapter 2 was published in the inaugural version of the journal *Consumption and Society*.[1] Chapter 4 is an expanded and modified version of a book chapter published in *Consumption, Status, and Sustainability: Ecological and Anthropological Perspectives*.[2]

To my family and friends who supported the fieldwork and writing of this book: Amma and Appa for inculcating a love of books and setting very high English standards (though you will be amused to know that I continue to mix up words all the time) and Appa for train tickets that got me to and from Bengaluru every few weeks; Sumi for all the help through the years, especially in clutch moments; Shilu Pati for keeping it chill and for being the OG busybody; Sudhar and Geetha for providing shelter in Bengaluru when an interview ran late and I was stuck miles from home; Sneha and Ameya for pensive companionship in Bengaluru bars; Anup who took me to Koshy's to eat pork when I was trying to shake off my vegetarianism; Sumithra for refuge, comradery, and good advice spanning decades; Puneeth for Bengaluru context and comedy; Nija for giving me a home-away-from-home in Oakland; Mahiya and Sanaya for insisting that work should wait and that I needed to play hide-and-seek ASAP; Genevieve for cooking wonderful meals in Mollans while I wrote; Mamie and Salah for keeping us company; Sarah, Kianoush, and Maël for the mountain hikes; Freyja, Traci, and Jennie for the laughs and irony; Rashmi, Esther, and Danielle for your unflinching faith in me; and most especially to Naïm, my partner, whose help, patience, kindness, and love has not only sustained the arduous process of producing this tome, but is also the anchor to my existence.

Introduction: Situating Sustainability in Bengaluru's Discards

"Stop! Stop stealing the milk packets!" shouted Kranthi, a thirty-eight-year-old tech executive at a multinational company in Bengaluru, India. She was screaming at three *pourakarmikas* (sanitation workers) who were hastily chucking high-value recyclables from a moving truck onto the tin roof of a scrap shop. I was in Kranthi's car on a Sunday morning in November 2012 as we followed a truck collecting recyclables from apartment complexes in her neighborhood. These weekly collection trips were the centerpiece of the zero-waste initiative that Kranthi had launched in KK Nagar, a residential neighborhood that emerged in the 2000s alongside a tech-industrial park in Bengaluru's southwest periphery.[1] Kranthi was upset because she saw the actions of the *pourakarmikas* as a betrayal of the initiative's ideals—to keep recyclables out of landfills by diverting them to "legitimate" value chains like the Wealth Out of Waste program run by ITC (formerly known as the Indian Tobacco Corporation).[2] Instead, the workers were subverting her green visions by claiming the value of the waste for themselves and doing what they had long done—selling recyclables to scrap dealers in the informal economy to supplement their meager pay.

"I wanted to invest some of my time and energy to improve the long-term health and sustainability of my community," Kranthi told me when I asked her why she had started this effort. Frustrated seeing garbage on her street day after day, she impugned the municipality and the waste contractor. "They kept not collecting garbage regularly. . . . I am quite sure he is corrupt." But when an activist from a citywide advocacy group made a presentation at her workplace about community-based waste management programs, she decided to take matters into her own hands. With the help of three retired men, Kranthi recruited two hundred households, asking them to segregate their waste into three categories: dry, wet, and sanitary waste. Wet waste was collected daily by the private company contracted by the municipality to collect waste in this area. The same workers collected dry waste once a week, transporting it to a waste-sorting center (a shed with no electricity), where it was weighed and sold to ITC for a bulk rate of 2 rupees (.03 USD) per kilogram.

During these weekly trips, Kranthi and her band of eager retirees followed the *pourakarmikas* to monitor how well KK Nagar residents were adapting their behaviors. At each of our stops, the workers, two women and one man, went through bins pulling out paper, plastic, and other recyclable materials. Kranthi asked them to put on gloves, but the thin latex tore in contact with coarse waste. One woman wore a jacket issued as a uniform by the contractor. Otherwise, there was no personal protective equipment in sight. The *pourakarmikas* bent and knelt, sorting on the floor, while we stood staring. Although the apartment complex residents were supposed to have pre-sorted the waste, segregation was imperfect. When we found rotting flowers and diapers in a bin marked dry waste, Kranthi immediately admonished the apartment-level project leader for the misstep. The workers moved the diaper to the correct bin. Soon, after visiting eight apartment complexes, we headed toward the sorting center, following the open-backed truck with bags of recyclable material. We lost sight of the truck, which appeared to take a turn into a side lane. Kranthi caught up, and as we came closer, we saw that the truck had stopped next to a scrap shop, and the *pourakarmikas* were rapidly throwing out the highest-value recyclables. Out went the cardboard boxes and the milk packets. Once the workers realized we had caught up with them, they stopped and quickly drove on.

At the sorting center, Kranthi confronted the workers about what happened. They sidestepped her censure with half-hearted nods, voicing their frustrations to their boss in Kannada: "We would get so much more money if we sold the *maal* [valuable materials] to the scrap shop than just giving everything to ITC for 2 rupees. It is stupid to give it away so cheap."[3] "Shut up," shouted their supervisor. Kranthi tried to placate them: "After all, we give you some of the money from the sale to ITC, and breakfast too." The *pourakarmikas* were unhappy with the scrutiny. Even if the money accrued by selling recyclables to ITC was ultimately distributed to the workers (after deducting unspecified "operating costs"), they resented losing control over the process. The middle-class volunteers, in contrast, had no visible financial motives and did not see the need to sell the recyclables for a higher price. Their priorities were clean streets and a clean ecological conscience. And they wanted to get back home to hot lunches, with an afterglow of a morning's worth of civic duty under their belt.

This conflict captures a key tension in urban environmentalisms in Bengaluru, India. On the one hand, well-meaning environmentalists like Kranthi desire clean and green cities, which they hope to achieve through behavior change, infrastructural upgrades, and market mechanisms. Some work hard in their homes and neighborhoods to execute community-based environmental solutions. But their attempts at achieving sustainable cities requires the labor of waste workers, who they retain in class- and caste-subjugated roles. Their zero-waste schemes also threaten long-standing circuits of resource recovery tied to informal livelihoods and economies. They do not readily recognize sanitation workers, waste pickers, and scrap dealers as legitimate environmental or economic actors, excluding them from full participation in urban environmental politics.

But within this story of oppression repackaged with a green sheen are potential seeds of change. Metabolic dependencies on labor to enact zero-waste systems had opened new avenues for environmental claims-making by certain sectors of the urban poor in Bengaluru. Indeed, a few weeks before my visit to KK Nagar, I was in the Karnataka High Court. In a crowded courtroom, two judges were presiding over a public interest litigation (PIL) filed by a group of housewives, retirees, and some nongovernmental organizations affiliated with a network called the Clean and Green Forum (TCGF),[4] accusing the Bengaluru municipal authority of violating their constitutional right to life by failing to manage Bengaluru's waste.[5] TCGF members went to the courts after several years of neighborhood-level advocacy promoting "zero-waste" practices.

Alongside TCGF in the courtroom were representatives from Parisara Tanda, a newly formed organization advocating for the rights of informal waste pickers—individuals, families, and communities who make a living by collecting, sorting, and diverting recyclable materials (Dias 2016). Waste pickers are "other low carbon protagonists" (Cohen 2017) who participate in the environmental management of cities and the maintenance of global environmental health, but on embattled terms. Since 2011, Parisara Tanda has worked to gain recognition and inclusion for waste pickers in Bengaluru, helping them access social services and economic opportunities in

emerging recycling systems. Parisara Tanda's organizing strategy emphasizes building alliances with middle-class environmentalists: "Our initial
goal was to get them [middle classes] comfortable with the idea of including informal sector and waste pickers. If not, they will give the contracts
to corporates. I have seen it everywhere; when middle class get involved in
waste, waste pickers get wiped out," explained one of the organizations'
founders in a 2012 interview. Leveraging the green talk of its middle-class
partners, Parisara Tanda reframed informal waste workers as the city's "silent
environmentalists" and "robust entrepreneurs," who recovered resources,
revalorized discards, and accumulated ecological knowledge in the shadows.

The PIL was well-timed. In August 2012, Bengaluru was plunged into a
garbage crisis when *pourakarmikas* across the city went on strike protest-
ing unpaid wages and poor working conditions. Compounding matters,
the residents of Mandur and Mavallipura, two peripheral villages that
received Bengaluru's discards, constructed blockades to stop garbage trucks
from dumping waste.[6] Years of concerted organizing by the Dalit Sangarsh
Samithi had finally brought media attention to the plight of the communities living next to these dumpsites. With the dumps closed, stinking
mounds of garbage piled up on every street corner. Rats and bandicoots
came out at night, frolicking in paper, plastic, and vegetable peels. English-
language newspapers and 24/7 news channels declared a garbage crisis. TV
journalists stood next to open dumps, portending public health disasters.
A *New York Times* article titled "India's Plague, Trash, Drowns Bengaluru,
Its Garden City" seemed to push the city over the edge, into straight-out
panic mode.[7] Sensing an opportunity, a prominent environmental justice
organization called the Environment Support Group filed complementary
legal petitions on the rights of landfill-adjacent communities and the need
for democratic, decentralized governance of waste infrastructures. Together,
this coalition of middle-class environmentalists, environmental justice
groups, and labor organizers demanded that the municipality mandate segregation of waste at source, build recycling and composting facilities, and
institute citizen committees to oversee these operations.

It worked. Facing mounting pressure and desperate for new solutions, the
judges did something very surprising. Instead of deploying police to reopen
the dumps and rejecting counterproposals to set up waste-to-energy facilities, they ordered the municipal government to stop dumping waste and
begin constructing a decentralized network of materials recovery centers in

Bengaluru.[8] Parisara Tanda also succeeded in gaining formal recognition of waste pickers when Bengaluru became one of the first municipal authorities to issue occupational identity cards to waste pickers. With support from key middle-class movement actors, Parisara Tanda had secured new economic opportunities for waste pickers to operate recycling centers and provide waste management services.

In a decade, waste pickers in Bengaluru have gone from personae non gratae, people to be avoided and expunged, to celebrated environmental heroes. Bengaluru is now hailed as a leader in informal sector integration and community involvement in waste management.[9] Several other Indian cities have issued occupational identity cards to enumerated waste pickers, and informal sector inclusion is emphasized in Narendra Modi's signature 2014 Swachh Bharat "Clean India" Mission.[10] Parisara Tanda's waste pickers partner with global brands like H&M and The Body Shop, have been featured in lifestyle magazines, and speak at national and global environmental summits. Sustainability and environmentalism became the terrain on which they defended their right to the city. This book tells the story of how this happened and asks what it can tell us about the potential for and pathways to *just sustainabilities* in global cities.[11]

A product of my decade-long engagement with Bengaluru's zero-waste movement, *Recycling Class* shows how diverse social groups adopt, contest, and modify neoliberal sustainability's emphasis on market-based solutions, behavior change, and aesthetic conflation of clean with green. Tracing the flows of both waste materials and sustainability discourses, I link an examination of middle-class (sustainable) consumption with the (environmental) labor of the working poor to offer a relational analysis of urban sustainability politics and practice. It is well past time to move beyond simplistic accounts of sustainability that frame the expanding Indian middle class as rapacious, environmentally unconscious consumers, or the urban poor in Global South cities as passive recipients of environmental injustice. In the discussions that follow, I offer an alternative perspective that instead analyzes consumers' and waste-pickers' interactions within urban waste metabolisms.

Based on ethnographic and community-based research, I demonstrate that the very presence of waste pickers in what I term *communal sustainability* challenges the existing discourse and forms of environmentalism, forcing the middle classes and the state to consider livelihood, occupational health, and social welfare as crucial elements of sustainability transitions.

While waste-picker organizations have used the waste and urban sustainability agenda to create new avenues for economic inclusion and political negotiation, the same agenda also reproduces unequal distributions of risk and responsibility. Drawing on feminist geography and urban political ecology, I argue that achieving just and ecologically safe cities ultimately requires resisting the seduction of neoliberal logics of growth, efficiency, and "clean and green" aesthetics to reclaim the city from the detritus of a consumerist and casteist society.

World-Class Cities Are Wasteful Cities

This book is set in Bengaluru, arguably India's most cosmopolitan city. A significant proportion of Bengaluru's ten million residents are "new middle class," the term used in sociological scholarship to describe India's emerging, globally connected elite. The hub of India's $150 billion information technology (IT) industry, Bengaluru encompasses unique demographic characteristics that make it "not-your-typical-apocalyptic-megacity." Instead, it is a city of public-sector enterprises, educational institutions, tech companies, organic boutiques, and English-style pubs. Its distinctive urban formations are sustained today by the one to two million IT and other professionalized workers who work in multinational and Indian technology corporations, investment banks, media, health care, and other service sectors. The lives and lifestyles of these high-tech workers are global, from their working hours, which are often in sync with US and UK time zones, to the kind of clothes they wear, the food they eat, the neighborhoods they live in, and the products they buy. Bengaluru is a critical node in a circuit that moves skilled migrant bodies through global tech hubs like San Francisco, Seattle, and Melbourne. Its elites are often the first to adopt what is in vogue in these global cities.[12]

My relationship with Bengaluru began as a child. My mother worked in the city as a journalist when she was pregnant with me, riding the public buses with her growing belly. I spent the first eight months of my life living in a railway colony behind Majestic, the city's busy transportation hub. We visited Bangalore (its colonial name) frequently when I was growing up in Chennai (previously Madras), another metropolis in Southern India. For Madrasis like me, these visits were a sensory delight. Chennai was loud, crowded, busy, humid, and frenetic. The parts of Bangalore I visited as a child, which included the tree-lined avenues of Malleswaram and the old groves in Ulsoor Lake, were cool, breezy, pleasant, and relaxing. We used to call Bangalore the air-conditioned city because of its temperate climes. I moved back to the city in my early twenties, seeking employment and a life unburdened by the traditional expectations of marriage and childbearing. I went to Bangalore because I found a job there, but also to escape parental and family scrutiny of my life choices. I drank in the city's many pubs (a taboo activity for women in India), started eating meat to the disappointment of my Brahminical family, jogged around Sankey tank, and enjoyed being a young consumer in a rapidly changing city.

Places like Bengaluru are viewed in contradictory ways; for growth boosters, they are sites of innovation and the future of tech-capitalism. The media and the government celebrate Bengaluru's cosmopolitan consumers as incontrovertible evidence of the efficacy of India's neoliberal economic reforms. But for environmentalists in the West, rising middle-class and elite consumption in Asia is a source of deep anxiety. From 2000 to 2015, the middle and upper classes of emerging countries increased their emissions more than any other group, reflecting increased incomes and access to global markets and consumer goods (Chancel and Piketty 2015). Although feverish portrayals of voracious consumers in China and India might whip US environmentalists into a frenzy, these broad-brush characterizations ignore the uneven nature of Asian development on the ground (Roy and Ong 2011). Indeed, the presence of high-consuming groups amid a sea of poverty reflects the profound inequities of India's economic successes (Oxfam International 2019), just as the high ecological footprints of Western consumers, even in comparison with high income groups in Asia, reflect the living histories of imperialism and racial-capitalism (Hickel et al. 2022). These forces concentrate wealth and comfort in privileged spaces and populations while extracting labor and resources from, and dumping waste in, the majority world (Patel and Moore 2018).

Imperial forces in their neoliberal avatars have also shaped Bengaluru's spaces and infrastructures. Starting in the late 1980s, structural adjustment programs imposed by the International Monetary Fund opened India up to foreign direct investment, deregulated a range of industrial and service sectors, and privatized public-sector enterprises (Fernandes 2009). These changes brought jobs and opportunities to dominant caste and middle-class urban Indians like me, while pushing others into further precarity. Whereas

symbols of globally connected lifestyles are abundantly visible in Bengaluru's swanky malls, chic coffee shops, gated communities, and car-clogged
roads, markers of poverty can be spotted behind the tall buildings: in
migrant worker camps, in the city's slums, and in its garbage dumps. Neoliberal policies that promote private consumption and underfund public
infrastructure have encouraged the city's elite to consume and waste in
order to power the engines of consumer capitalism, while relegating slum
dwellers to intermittent water and energy supplies. Migrants sustain themselves by scavenging the city's discards, and COVID-19 patients across social
classes used Twitter to hunt for hospital beds and oxygen cylinders during India's devastating Delta wave. But the city's, nay nation's, self-image
depends on hiding away these ruptures and failures, and projecting itself
as a vibrant, innovative, and investment-worthy "world-class" destination.

Bengaluru today is a different place than the cool oasis I imagined it to be
in my childhood. The IT boom has cleaved and expanded Bengaluru. State-
designated special economic zones cut through grazing lands and agrarian
villages. Massive gated complexes, many built on ecologically sensitive lands
commandeered for real estate construction through shady land deals, encircle
the city, housing both tech workers and global capital seeking productive
sites of investment (Doshi and Ranganathan 2017; Goldman 2011). Criss-
crossing networks of elevated expressways dominate the landscape, plying
the thousands of automobiles that are added to the roads every year (Gopakumar 2020). Critics argue that these changes have made Bengaluru a dys-
topian nightmare (Menon 2017). A city that was once narrated as India's
garden city or a "pensioner's paradise" of tree-lined boulevards and colonial
neighborhoods is now ranked as India's least-livable large metro. Scientists
at the Indian Institute of Science warn that flooding and water scarcity will
make the city unlivable as early as 2030. Bengaluru's lakes are burning and
spewing chemical froth. Heavy rains in 2022 inundated many peri-urban
areas, sparing neither the homes of CEOs nor the informal settlements
where sanitation workers and waste pickers live (Ranganathan 2022). The
city oscillates between droughts and floods. Despite these warnings, the
juggernaut of eviscerating urbanization moves on, punctuated by cyclical
garbage crises.

Garbage is a flashpoint for debates about urban futures in Bengaluru, just
as it is in many majority world cities.[13] The city's development model aims
to create a "clean and green," modern, globalized city of consumers, but this

very model produces ever-escalating amounts of daily discards that it struggles to cope with. Bengaluru produces somewhere between 5,000 and 6,000
tons of solid waste every day, a number that has quadrupled in two decades.[14]
As cities like Bengaluru have grown in size, population, and GDP, what gets
thrown away has changed. There is more plastic in municipal waste, much
of it nonrecyclable packaging materials used in those fast-moving consumer
goods that urban Indians buy in growing quantities, as well as more construction and demolition waste from real estate development. These materials do not biodegrade, a property that renders open dumping, India's
long-standing primary mode of waste disposal, woefully inadequate.[15]
Waste dumping subjects the urban poor and peri-urban communities to a
slow but deadly violence.[16] In the grazing and agrarian communities that
receive Bengaluru's discards, groundwater is slowly poisoned. Children contract malaria and dengue from mosquitoes thriving in standing cesspools.
Sometimes, people die, as they did in Mavallipura in 2012, prompting its
residents to use their bodies to block garbage trucks.

Neoliberal agendas drive urban governments to pursue market-based and
technocratic approaches to solve environmental problems. Bengaluru and
other Indian cities have responded to cyclical garbage crises with the same
neoliberal playbook, privatizing service provision, casualizing labor arrangements, and pursuing tech fixes that are poorly suited to local conditions,
such as waste-to-energy plants (Luthra 2015). These interventions diagnose
infrastructural deficits as a product of poor state practice, as opposed to lack
of resources or structural inequalities (Doron and Jeffrey 2018). Moreover,
although neoliberal programs like "smart cities" often purport to empower
citizens and increase participation in municipal governance, they subvert
democratic processes and can lead to splintered infrastructural development.[17] In Bengaluru, privatization strengthened the "garbage mafia"—
private contractors from land-owning castes for whom waste management
is a lucrative business—while worsening service provision for urban residents, labor conditions for sanitation workers, and waste access for informal waste pickers.

Suturing together patchy "formal" waste infrastructures are a network
of waste pickers, sorters, itinerant buyers, scrap dealers, and recyclers. This
parallel informal economy is based on extracting recyclable and reusable
material from dumped mixed garbage, creating economic value (Gill 2009).
At its core are waste pickers, millions of people across Asian, African, and

Latin American cities who make a livelihood reclaiming value from waste—some seeking more autonomy, some to escape exploitative waged work, and others seeking any possible employment in cities where work is hard to find (Dias 2016; Millar 2018).[18] This informal system provides an environmental service to the city by extending the useful life of commodities and diverting recyclable materials from dumpsites, while also reducing waste transportation costs for the municipality.[19] Yet, these "vernacular" circular economies are often denigrated as polluting and inefficient, and targeted for reform or replacement (Tucker and Anantharaman 2020).

In Bengaluru, as with other South Asian cities, working with waste is a stigmatized occupation because of its relation to caste. Waste removal through manual scavenging has been forced on Dalits, historically referred to as "untouchables," a term I henceforth avoid because its very use is a form of symbolic violence (Shankar and Swaroop 2021). Caste enshrines inequality and dehumanization in routine ways in Hindu society (Ambedkar 2014; Guru 2011). Dalits and others who work with waste, especially human waste, are routinely subjected to physical, structural, and symbolic violence (Doron and Jeffrey 2018; P. Gupta 2022). Working with waste is associated with humiliation, while also exposing pickers to toxic substances and deadly working conditions. Those who work with waste are usually assumed to be lower-caste or Dalit, thus subjecting them to caste-based stigma (Kornberg 2019a). In other cultural contexts, waste picking is racialized work, disproportionately carried out by Afro-Brazilians, Indigenous people, and ethnic and religious minorities.

This is a familiar story not just in Bengaluru, but globally. Colonial, racial capitalism distributes hazards disproportionately onto marginalized communities that function as environmental sinks and subjects certain social groups to toxic and deadly work (Liboiron 2021). These processes, concentrating harm in certain places and bodies, operate at local, regional, and even global scales—indeed, Bengaluru is also the destination for e-waste discarded by consumers in the West, processed in its informal settlements by Muslim and Dalit recyclers (Reddy 2016). Meanwhile, environmental privilege allows people with economic and social capital to remain oblivious to these slow-moving, protracted disasters.[20] That is, unless and until they spectacularly erupt, whether via a strike by sanitation workers, or a blockade by the villagers of Mavallipura, or when China put up the "national sword," leaving US municipalities scrambling for destinations for their plastics. Disposability is an unfinished process (Liboiron and Lepawsky 2022). There is no real "away" for waste. These critical events, wherein garbage exceeds its designated boundaries and becomes matter out of place, threaten social order. They also illuminate flows, relations, and dependencies obscured within colonial-capitalisms' ecologically unequal systems of exchange, shattering mythologies and sometimes sparking new mobilizations.

Sustainability as Relational Terrain

This book takes the ideas, histories, flows, and relationships around waste in this critical node of global capitalism, itself a massive and growing environmental problem of planetary proportions, to push our understanding of what types of coalitions and movements are necessary to pursue just sustainabilities. My analytical approach is relational, situated, and plural. A relational approach, as practiced by situated urban political ecologists, sees individual subjectivities and the social world as (re)produced and transformed through everyday, materially mediated interactions among actors positioned differentially in social milieus (Lawhon, Ernstson, and Silver 2014). Such a perspective helps me argue that changes in environmental subjectivities are not simply produced by exposure to ideas from elsewhere, as presupposed by the predominant norm diffusion model of environmentalism emerging from theories of modernization and enlightenment. Ideas matter, yes, but environmental subjectivities emerge from everyday friction, obdurate material conditions, political struggles, community experimentation, and unlikely alliances.[21]

Sustainability is perhaps the most ill-defined word in the canon of environmental action (Greenberg 2013). It is an empty signifier par excellence. This very amorphous, yet increasingly ubiquitous quality makes sustainability discourse and practice a potent vehicle for politics, despite its apolitical and technocratic framing.[22] Like economic development, it is quickly emerging as a dominant ideology and framework for material action in global cities (Castán Broto and Westman 2019; Bulkeley 2013). It is also a disadvantageous terrain of struggle for the urban poor. Following the Relational Poverty Network's call to expand and enliven poverty research into new domains, I examine sustainability as a terrain or "contact zone" where "poor others" are engaged and acted on by powerful groups in the city.[23] At the same time, I show how the former resist, co-opt, and negotiate elite environmental activism. Ultimately, I argue that it is through "contact," occurring in this case through the material, metabolic bonds of garbage, that sustainability can be reclaimed as a progressive discourse that informs action on the ground.[24]

A relational approach also sees power as diffused and enacted through socio-material landscapes (Castán Broto and Calvet 2020). It moves away from reductionist, dualist conceptions of conflict, instead seeking nuanced explanations of power differentials and leaving open room to see how marginalized groups resist and negotiate environmental injustices (Le Billon and Duffy 2018). Situated analysis can help locate agency and reveal how power can be reclaimed in incremental steps through recursive empowerment (Pieterse 2021). It helps us look for seeds of change in exploitative systems and consider how transforming an unsustainable status quo requires engagement with it (Castán Broto and Westman 2019, 64). Thus, I begin with local context, identities, and everyday practice, and use epistemologies beyond those of the Global North to explore and explain the actually existing ecologies of the city. This approach is especially important because theories emerging from Western democracies do not always make sense in other places. For example, calls for more citizen and community involvement to expand the frontiers of urban sustainability do not travel well to cities where elites have captured most participatory initiatives, and where civil society is constructed to systematically exclude subaltern groups.[25]

Finally, I refuse to see diverse schools of environmental thought as incompatible.[26] Trained as an interdisciplinary scholar, I am a promiscuous intellectual. I combine the insights of Marxist geographers, political ecologists, cultural sociologists, discard-studies scholars, and theorists of caste and racial capitalism. I pay attention to social locations, material conditions, governing regimes, ideological orientations, and actions, in relation to one another. My analysis also considers state power, in its different forms, to explain why groups take up certain environmental strategies, when they succeed, and why they fail.[27]

Because of my situated and relational approach, this story might sometimes seem contradictory. I recount instances where the middle classes and the working poor have formed positive alliances, but then go on to describe how those alliances now make waste pickers beholden to a new set of interests and actors. I complicate the idea that economic inclusion into sustainability infrastructure is beneficial to informal workers by showing how inclusion is designed to primarily serve the ends of capital accumulation and neoliberal austerity. Ultimately, I argue that, despite opening them to new forms of exploitation, partnership with middle-class actors also gives waste pickers and their allies access to new arenas for political negotiation. Cross-class alliances around waste are contradictory, both challenging and reinforcing gender, class, and caste hierarchies. These contradictions are central findings of this work; scholars seeking just sustainabilities need to be attentive to both the oppressive and liberatory outcomes of particular events. As discard-studies scholars Max Liboiron and Josh Lepawsky assert, "As researchers, we must also account for permanent toxicity, gross inequalities and power differentials . . . a need to offer alternative practices as much as critique, and humble narratives that leave for open and diverse futures" (2022, 30). It is in this spirit that I offer this analysis.

In the three sections that follow, I highlight the theoretical contributions of this project to scholarship on environmental sociology and politics, global environmental justice, and sustainability studies, before discussing my methodological approach.

Everyday Environmentalism in Unequal Cities

My book contributes to *environmental sociology and politics* by critically examining the green consumption and community mobilization of Bengaluru's elite middle classes in relation to global sustainability discourses and the environmental labor and political claims of the working poor. Discussions of environmental issues in Asia are often framed as large-scale problems of state and global governance linked to the development of spectacular infrastructures or market mechanisms. But unstable urban ecologies are birthing diverse environmental social movements. While robust debates exist on the *environmentalism of the poor* of rural communities in South Asia, ethnographic portraits of urban environmental subjectivities are still emerging (Rademacher and Sivaramakrishnan 2013). Yet, South Asian cities are the densest and most-populated urban agglomerations in the world, housing about 700 million people, 130 million of whom live in informal settlements.

As South Asian cities strive to emulate "world-class" models like Singapore, tensions among growth, equity, and ecology regularly ignite around slum removal (Doshi 2019; Ghertner 2012), river and lake restoration (Rademacher 2011; Sen, Unnikrishnan, and Nagendra 2021), green space

development (Coelho 2020), food systems (Frazier 2018), streetscapes (Anjaria 2009), air pollution (Véron 2006), and garbage (Luthra 2018; S. Sharma
2022). Across diverse cases, the new middle classes, thanks to their growing
economic power and cultural hegemony, have had the most success gaining recognition for their environmental claims. Middle-class narratives of
environmentalism are extremely popular, especially among those who hold
power within the status quo. But sociologist Amita Baviskar cautions that
middle-class environmental efforts often have little to do with either ecology
or justice (Baviskar 2019b). Acting often through the judicial system, these
bourgeois environmentalists use discourses of hygiene and public nuisance
to refashion public space to match their aesthetic preferences, advocating
the demolition of informal settlements and displacement of informal livelihoods. This brand of exclusionary environmentalism is particularly attractive to the state because it aligns with capitalist, modernizing agendas.

But middle-class environmentalism is evolving. Postcolonial frameworks
of environmentalism offer an increasingly incomplete picture of practice-
based movements emerging under fragile infrastructures, neoliberal (self)gov-
ernance and austerity regimes, rapid local environmental degradation, and
circulating narratives of modernity in global crisis. "Everyday" or "lifestyle"
environmentalisms are proliferating across the Global North.[28] Bengaluru,
sitting at a global crossroads with a significant proportion of dominant-caste
diasporic returnees, yet embedded in local caste relations and ecologies, has
evolved an articulation of diverse environmentalisms together.

Responding to the degradation of their local environments, desiring to
keep up with counterparts in other global hipster cities like San Francisco,
and seeking to enact cosmopolitan ideals of ecological citizenship, sectors
of Bengaluru's elite middle classes are adopting a range of green lifestyle
practices. Many are riding bicycles to work, eating organic food, and recycling and composting their wastes. They convene communities of practice
to sustain and expand green lifestyles in the city. These everyday environmental movements are not unique to Bengaluru. Across Asia, experiments in green living are proliferating, some grassroots efforts and others
prompted by state schemes and green marketing (Lewis 2016). Members of
these communities talk eloquently about the importance of voluntary sim-
plicity, individual obligation, and ecological citizenship, a marked contrast
from previous expressions of elite environmental activism in Indian cities

that generally sought to evade environmental responsibility (Bulkeley and Castán Broto 2014; Frazier 2018). Instead, the mantras of these new movements include "it starts at home" and "eat what you grow."

These lifestyle movements and community-based sustainability efforts nevertheless reproduce class and caste oppression through aesthetic and discursive markers that delegitimize the knowledge, contributions, and practices of the poor, while simultaneously relying on them for metabolic labor to do green practices. In chapter 1, I draw on the cultural sociology of Pierre Bourdieu (1984) to retheorize the term *performative environmentalism* to describe environmental engagement through lifestyle changes and neighborhood activism by Bengaluru's cultural elites. Performative environmentalism is distinct from bourgeois environmentalism in that it does not deflect blame for urban dysfunction onto the poor. Instead, it questions urban growth trajectories and promotes ecological citizenship through collective efforts at changing consumption practices. It nevertheless *others* the poor through class- and caste-based aesthetic and discursive markers. Even though performative environmentalists might not actively call for slum removal, they employ *defensive distinctions* to distance themselves from the necessity-oriented green practices of the poor, in order to legitimize environmental actions within their own class and caste groups. In doing so, they exclude the urban poor from their conceptions of green community.

This concept can help us critically examine the cultural politics of everyday environmentalism in other unequal cities harboring diverse social groups (i.e., most cities). Cities around the world are investing in green infrastructures like public transit systems, bike lanes, and farmers markets as both climate and urban competitiveness strategies (Rosan and Pearsall 2018; Stehlin 2019). Yet, as *critical sustainability* scholars warn, actions carried out in the name of sustainability can burden racialized others and displace working classes that lack consumption power (Anguelovski and Connolly 2021; Isenhour, McDonogh, and Checker 2014). My concept of performative environmentalism adds to this literature by showing how well-meaning environmentalists (often inadvertently) participate in these exclusions through their cultural strategies of consumption. Performative environmentalism is sincere, ethically driven, and involves material action by committed individuals. However, at the scale of social structure, it functions as ideological gloss, reproducing class and caste hierarchies.

The scholarship on sustainability in Global North cities critiques how neoliberal ideologies have individualized and depoliticized environmentalism (Maniates 2001). As geographer Eric Swyngedouw explains, neoliberal post-politicization has created important roles for technocrats, business managers, and consumers while marginalizing other, more conflictual subject positions.[29] Only convivial and conciliatory community initiatives are sanctioned within neoliberal sustainability.[30] Yet, while these theorizations explain why environmentalism functions as anti-politics in Global North cities, they do not adequately capture what I witnessed in Bengaluru.

Bengaluru's circumstances compel political contention and coalition building as part of environmental projects.[31] While middle-class people might have hoped to get away from the state and the poor to meet their (greening) needs, the lack of market provisioning of green living and loose infrastructural threads exacerbated by neoliberal disinvestment necessitated community organizing. Elite and middle-class groups have long used civic associations like residents' welfare associations to make political claims on the state, whereas here I describe place-based material enaction of environmental politics. In chapter 2, I draw on intersectional theories of social reproduction developed by feminist geographers to articulate the concept of *communal sustainability*. Communal sustainability describes neighborhood-based interventions designed to replace spectacular infrastructure, technocratic expertise, and waste dumping. Instead, housewives, retired men, and other unlikely suspects deploy affective and reproductive labor to change household behavior, build small-scale infrastructures, and convene collaborative systems of governance.

Communal sustainability goes beyond simplistic explanations of environmental harm that blame lack of knowledge, bad behavior, or immorality for garbage crises to instead identify collective pathways to overcome shared environmental problems. But in its material solutions to environmental problems, communal sustainability mobilizes metabolic divisions of community that are gendered, classed, and casted. Originating in the caste system, the notions of community operational in these initiatives are restricted to the middle class and dominant castes, casting waste workers as noncommunity members who need to be disciplined and monitored (as in the opening vignette).[32]

At the same time, I caution that there are limits to seeing the performance of gendered and casted reproductive labor solely through the lens of governmentality or the extraction of surplus labor. Also operative here is a sense of empowerment, a building of shared identity, and an enactment of politics for those engaged in this work, which cannot be reduced to a narrow economism or top-down governmentalization. Rather, communal sustainability, in its dependence on volunteer effort and manual labor, challenges neoliberal sustainability's individualizing and marketizing tendencies. I highlight a paradox—communal sustainability can exacerbate existing class, caste, and gender-based oppression while simultaneously opening up new avenues for political participation by middle-class women and, as we will see next, by some waste workers.[33] In the discussion that follows, I center the unheralded *entrepreneurial environmentalisms* of the city's waste pickers, whose participation in zero-waste programs helped reframe garbage as a sociopolitical problem and not simply a technical, managerial, or behavioral one.

New Environmentalisms of the Poor

My book contributes to studies of *global environmental justice* by analyzing how waste labor becomes a route for oppressed groups to participate in sustainable city-making. The role of work and workers remains underexplored in the environmental literature. But, as geographer Rosalind Fredericks (2018) argues, analyzing urban labor as the basis of citizenship is a powerful lens through which to make "theory from the South" about the neoliberal era. Such theory-making pushes back against a reading of neoliberalism as a juggernaut, instead bringing attention to the multiple and contradictory ways in which diverse actors resist, appropriate, and transform neoliberal (environmental) ideologies and neoliberalizing processes.

The power to define what counts as sustainable is unequally distributed. Poor, working-class, and nondominant-caste people in Indian cities have struggled to gain recognition for their environmental claims (Baviskar 2019a). Because working-class urban livelihoods are more often associated with industrial or service activities, struggles to preserve these livelihoods or establish a "right to the city" are less readily legitimated as the environmentalism of the poor.[34] At the same time, because of their limited consuming power and consistent disenfranchisement in civil society spaces, these communities have fewer opportunities to perform environmentalism through consumption and civic activism.[35] Critically, because they are forced to live in "dirty places," which are dirty because the state has failed to provide basic infrastructures of waste collection, their habitat concerns are sometimes automatically deemed to be anti-environment by the state (M. Sharma 2017b). In extreme cases, they are simply not recognized as legitimate citizens of the urban polity (Chu and Michael 2019).

Exclusive and antipoor environmentalisms perpetuate Band-Aid solutions. As long as the poor and working classes are scapegoated for the problem of waste, for instance, plastic producers, brands, and the state can avoid critical attention and censure. This was certainly the case in Bengaluru, where for a long time, middle-class environmentalists blamed sanitation workers for the failure of recycling initiatives. Informal waste pickers and recyclers, the only agents recycling and recovering resources in Asian, African, and Latin American cities, have long operated in the shadows, evading punitive municipal policies and paternalistic formalization projects (Dias and Samson 2016).[36] Even eco-activists like Kranthi were ignorant or distrustful of informal waste economies, preferring to partner with private companies like ITC. Despite the antipoor nature of environmentalism, many members of the urban poor have no choice but to participate in the sustainability conversation, even if on unequal terms. Not engaging might only bring further violence and dispossession.

Tracking the formation, evolution, and activities of Parisara Tanda for over a decade, I document a new expression of the environmentalism of the poor. Parisara Tanda, which translates to "green force" in English, was informally founded in 2011, just as the middle-class zero-waste movement (ZWM) was recruiting corporate partners to run recycling centers in the city and enclosing recyclables within privatized value chains. It was created by middle-class labor organizers and community workers with the explicit goal of halting the corporate capture of Bengaluru's nascent zero-waste systems. Parisara Tanda, learning from the playbook of waste-picker organizations in South Africa, Brazil, Argentina, and Indian cities like Pune, built a collective identity for waste pickers by emphasizing the environmental and economic contributions of their work. But in contrast to these other cases, Parisara Tanda prioritized building strategic coalitions with middle-class environmentalists and approached the state in partnership with middle-class organizations.[37] To convince the casteist ZWM of waste pickers' worth, Parisara Tanda engaged in what one scrap dealer in a 2013 community meeting called a "face-lift," refashioning the public image of waste pickers and scrap dealers from abject and dangerous to *entrepreneurial environmentalists.*

While *global environmental justice* scholarship has focused largely on studying and defending the environmental rights of urban populations, I show that the "gospel of eco-efficiency" (Martinez-Alier 2003), central to neoliberal environmentalism, has influenced the environmentalism of the poor. In chapter 3, I rely on postcolonial theorist Gayatri Spivak's (1999) concept of strategic essentialism to explain how middle-class sustainability discourses enable new modes of claims-making for subaltern groups in the city. I deploy the phrase *of use to the city* as a counterpoint to *the right to the city* to discuss how Parisara Tanda uses neoliberal discourses to protect waste-based livelihoods. It leverages performative environmentalism's aesthetic and behavioral biases, alongside neoliberal discourses of entrepreneurial valor, to make political claims for waste pickers' rights and urban belonging.[38]

By examining sustainability not just in discourse, but also through practices of discarding, reclaiming, and recycling, I advance new concepts that consider waste infrastructure as sites of infrastructural citizenship by both elites and excluded groups. While infrastructure is commonly understood as physical provisioning systems comprising roads, pipes, and wires, urban political ecologists have drawn attention to how people, through practices and social relations, function as social infrastructures enabling cities and economies to function (Furlong 2011; Simone 2004). Waste and water infrastructures are also the sites of political action, through which excluded publics make themselves known, seen, and heard.[39]

Doing sustainability in Bengaluru evokes *infrastructural citizenship*, from the elite who seek to fashion solutions that materialize their environmental sensibilities, and from the poor defending their livelihoods. Elites and middle classes exercise undue influence over urban infrastructures. Participatory mechanisms convened by the Indian state have elevated the voices of elite and educated citizenry in urban governance regimes that value managerial capacities over labor.[40] Despite the odds, informal waste pickers and recyclers also influence the form of sustainability infrastructures, leveraging both their labor and their accumulated knowledge of waste materials to facilitate flows and reveal ruptures. Metabolic relations around waste are where these groups resist, co-opt, and negotiate each other's conceptions of problem and solution.

In chapter 4, I show how the entrepreneurial capacities of informal waste pickers combine with the aesthetic and civic sensibilities of the middle classes to coproduce what I call *DIY infrastructures*. A cash-strapped municipal government supports DIY infrastructures because they transfer the responsibility for urban sustainability and service provision from the municipality onto waste pickers and middle-class women. DIY infrastructures advance more democratized and participatory forms of sustainability, challenging the technocratic capture of waste management systems. Yet, its efficacy as a technology that "discards well" is undermined by scalar mismatches and sociocultural hierarchies that enable elites to pass off the burden of operating infrastructure to less powerful groups. Waste pickers in turn leverage their inclusion into these infrastructures to exercise citizenship. They engage in *materially mediated activism* to redesign systems and policies. Because of their advocacy, in 2018, the Bengaluru municipality decided to issue dry waste collection contracts exclusively to waste pickers and provide them with collection vehicles. This *stealth remunicipalization* of waste infrastructures was a significant change from the trajectories of privatization that Indian cities have seen since the early 2000s. Although these gains are contingent and progressive policies are constantly under threat of being reversed, they demonstrate that waste-picker organizations are a political force shaping urban futures.[41]

Ultimately, I demonstrate that global urban sustainability, with its pressures to green the city, green lifestyles, and, through this, display urban vitality and competitiveness, has created new motifs and avenues of social inclusion, exclusion, and political participation. Today, waste pickers are leveraging environmental arguments to make political claims not just in Bengaluru but also in South Africa, Argentina, Brazil, at UN Climate Summits, and, most recently, in the negotiation of the global plastics treaty.[42] In doing so, they are expanding the notions of who gets to participate in local and global environmental negotiations. Waste-picker organizations have made successful claims for inclusion into municipal waste infrastructures in Johannesburg, Pune, Belo Horizonte, and other cities. At the same time, inclusion is a double-edged sword. Governments engaging in "roll-out neoliberalism" are constantly conscripting (already oppressed) communities into conducting basic metabolic functions to keep cities functional, at lowest cost (H. Anantharaman 2019). As sociologist Manuel Rosaldo (2019) points out, waste pickers navigating unjust urban regimes often must choose between dispossession and exploitation (see also O'Hare 2020). Next, I evaluate the outcomes of these claims-making strategies and infrastructural collective action, asking both what they do for waste pickers and what we can learn from them.

Toward Cross-Class Collective Action for Just Sustainabilities

My book links scholarship on *just sustainabilities* and *discard studies* by examining how ideas about sustainability disrupt, dismiss, and change the lives of those who depend on circuits of waste for their livelihoods and, in turn, how waste workers alter urban sustainability agendas through their labor and activism. Just sustainabilities scholars Vanesa Castán Broto and Linda Westman (2019) assert that to recover sustainability from its techno-managerial shackles and make it a force for social justice, we need collective action that starts from the situated experiences of urban citizens, centers the experiences and needs of oppressed groups, and cultivates leadership from below. Yet, gross inequality impedes collective action on environmental issues, both because how people experience, understand, and respond to environmental problems varies based on their social locations and because these differential positions indicate different relationships to and perspectives on systems of oppression.[43] In a world where the oppression of some maintains "good" environments for others, sustainability projects shape and are shaped by social location (see also Sze 2018). Therefore, following these insights of intersectionality and diversity theorists, I assert that to identify pathways to just sustainabilities, studies of environmental collective action must consider sociocultural inequalities, racialized social relations, and identity formation to evaluate the emancipatory possibilities of different collective action projects.[44] I explore these dynamics among Bengaluru's discards.

How people problematize waste as an environmental problem is situated and partial. For oppressor caste, middle-class consumers, caste and class privilege keeps them at the center of urban life, from which waste is discarded away onto peripheral "sacrifice zones" (Lerner 2012). Caste blindness and denial enables (often willful) ignorance. Add neoliberal narratives that bad consumer behavior and poor disposal infrastructure are the driving causes of garbage crises, and you have a powerful mythology as to why the city is dirty and what can be done about it. A large and coordinated network of big brands and fossil fuel companies maintain this mythology by promoting piecemeal recycling schemes that circumvent regulation, buttress corporate profit, and dovetail with a green-growth agenda (MacBride 2011; Mah 2022). Environmental campaigns and scientific research on plastic and e-waste pollution place the onus for fixing the waste problem firmly onto consumers and municipalities.[45] Industry promotes recycling because it is profitable to move the costs of dealing with disposables onto the public, while also giving the impression that disposables are a sustainable type of waste (Liboiron and Lepawsky 2022, 70). Yet, many materials marked as recyclable are often not, because of lack of technology, fluctuating prices, or contamination.[46] Nevertheless, self-avowed environmentalists like Kranthi devote themselves to their local schemes, blaming the poor when their misguided efforts fail. The mythologies of neoliberal sustainability keep them trapped in these false forms of environmentalism.

Problem framings determine action: the stories people tell about what drives socio-environmental issues shape how they respond to them. Thus, it is imperative to ask, what forms of environmental collective action reinscribe existing mythologies that see pollution not as a constitutive feature of a colonial-capitalist economy, but as a mere externality that can be eliminated or displaced with the right markets, technology, design, and behavior? In contrast, what forms of collective action enable "defamiliarization and demythologization,"[47] that is, the transformative learning that changes how waste is problematized as an environmental issue, and births new environmental subjectivities and movements that center issues of justice? The final two chapters of this book contrast two forms of environmental collective action to explore these questions.

In chapter 5, I examine how urban struggles over the environment and livelihood articulate with transnational sets of questions around how to restructure global capitalism in a time of overlapping ecological crises. I examine transnational circuits of resource recovery articulating around the *circular economy*, which connect multinational corporations, global development actors, start-ups, and waste-picker organizations. Accelerating fears of resource shortages and the search for new productive frontiers of economic growth combined with heightened awareness of how waste and plastic pollution contribute to climate change and biodiversity loss have intensified efforts to reclaim value from discards. Multinational brands are looking to include waste pickers in recycled plastic supply chains, recognizing their unique knowledge of waste materials and needing cheap labor. Start-ups function as *legibility brokers*, linking waste pickers with the circuits of global capital. I find, however, that the "win-win" narrative of the circular economy is turning inclusion into a Trojan horse for capitalist gain.

The win-win narrative of a circular economy as applied to informal waste pickers obscures a tension between corporate profits and material improvements to the lives of informal workers. Circular economy discourse is replete with neoliberal "common sense" that promotes economic efficiency and market-based solutions to environmental problems. These logics subject waste pickers to new regimes of discipline that reproduce a dehumanizing division of labor. These initiatives can help some entrepreneurial waste pickers enhance their skills, improve incomes, and imagine social and career mobility for themselves and their children, but these benefits are not available to most. When initiatives privilege economic growth and corporate profit over other social priorities, *accumulation by inclusion* is the outcome. Ultimately, reclaiming value from waste in support of "win-win" sustainability goals (whether the clean city or circular economy) enshrines the continued wasting of people by retaining them in subordinated roles. When workers are held in subordinate positions and framed as needing to discipline and improve themselves, they have fewer opportunities to function as political or pedagogical agents.

Inclusion is a buzzword in sustainability and circular economy discourse today. The inclusion of waste pickers into resource recovery and recycling schemes has finally become a priority for global development organizations, the UN system, and for some national governments. In the coming years, as national governments, social movements, and multinational corporations negotiate the global plastics treaty and the implementation of the Sustainable Development Goals, Bengaluru's story cautions that the reliance on caste-determined environmental labor to produce sustainable cities or circular economies ultimately serves the ends of capital accumulation. Racial capitalism infiltrates and articulates with environmental mobilizations in contemporary cities, conscripting even well-meaning environmentalists into facilitating repackaged forms of exploitation. My work issues a warning: inclusion in the absence of a clear commitment to social reform and reparation could very well reproduce racialized oppression.

Yet, garbage is democratizing. It presents a moral and ethical problem that invites individual and community action to address its proximate causes and impacts (Douglas 2003; Hawkins 2001). In cities across the majority world, waste has played an outsized role in sparking civic activism and environmental mobilization, sometimes toppling governments and igniting social revolutions (Loschi 2019). In a more communal vein, civic-minded individuals encourage their neighbors to carefully segregate their discards, hoping to reduce the amount of waste ending up on the roadside and in landfills. Workers toil in materials recovery facilities, surrounded by piles of paper, plastics, and metals. Waste infrastructure, through the people it brings together and metabolic connections it creates, can also provide opportunities for political resistance, and even convene new relationships and unexpected coalitions.[48] Within communal sustainability initiatives, identification with a common set of goals opens new avenues for collaboration between middle-class women and waste workers. Building of shared identities and common projects like DIY infrastructures can build community assets, sustain cross-class collective action, and even result in shared political demands (Green and Haines 2015; Rigon and Castán Broto 2021). Recycling programs are not just vehicles for corporate greenwashing, but also domains for the expression of infrastructural citizenship and claims-making. Thus, despite opening themselves to new global circuits of exploitation, collaboration with middle-class environmentalists gives waste pickers access to new arenas for political negotiation through infrastructural citizenship.

In the conclusion, I take inspiration from the ways in which waste pickers and their representatives have engaged in a concerted program of popular education. Waste-picker organizations, through a perceptive use of traditional and social media, chronicle the doublespeak of the state and the failures of performative environmentalism. They reclaim political agency in communal sustainability through their work and advocacy at reflecting back to the city the metabolic flows it attempts to hide. The storytelling that waste pickers and their advocates do in these spaces is an act of radical pedagogy. It demystifies recycling as an industrial process, demonstrates the limits of downstream solutions to plastic pollution, and is forcing middle-class actors to problematize throwaway culture and consumer capitalism more centrally. Yet, for justice to be achieved, environmentalists must also grapple with the ways in which their initiatives condone and perpetuate casteism.

In closing, I argue for an agenda of *reparation by inclusion*, which leverages sustainability and circular economy to demand more material and intellectual investment into informal economies, as well as access to more social entitlements for informal waste pickers. It resists rebranding self-exploitation as entrepreneurism and deemphasizes efficiency in place of justice. It starts with a recognition that informal economies have long subsidized the conditions of capital accumulation, and that any sustainability effort will have to acknowledge and redress existing injustice through grassroots leadership.

Enacting reparation by inclusion will require the circular economy and other sustainability paradigms to go beyond a narrow economic rationality to reemphasize mutual vulnerability through transformative coalitions.

This book ultimately argues that by bringing together unlikely alliances anchored in material action, communal sustainability serves as a terrain that can produce politics that impede neoliberal capture. However, this transformative potential of local action is constantly threatened by the commodifying and co-opting forces of global capitalism, which threaten to subsume place-based experiments into a corporate green-growth agenda. It is also compromised by the denial of casteism, which locks waste pickers into subordinate roles in communal schemes. Even though communal sustainability, as emerging from and practiced by middle-class zero-waste warriors, politicizes the waste crisis by connecting it to urban development and consumerism, it does not question caste, and thus struggles to emerge as a transformative coalition centering justice and equity. Communal hierarchies and differences (along lines of race, caste, and ethnicity) undermine collective and communitarian routes to just sustainabilities. To move forward, privileged environmentalists, scholars, and activists must listen to waste pickers and prioritize social reform and reparation over aesthetics or efficiency.

My Relation to This Project

The methodological approach of this work is long-standing and community-engaged, focused on understanding everyday life through participant observation, discourse analysis, and qualitative interviews. This book narrates the making of a sustainable city through everyday practices within homes, in neighborhoods, in planning and advocacy spaces, in recycling centers, and on the street. With a decade of concerted engagement with communities and individuals involved in Bengaluru's zero-waste movement, I offer a long-term perspective. My work with Parisara Tanda, the waste-picker organization profiled in the book, offers a unique vantage point from which I can examine the politics and outcomes of sustainability initiatives. I offer more explanation about the data and fieldwork for each chapter in the accompanying notes.

The bulk of the ethnographic work on this project took place from 2011 to 2013. I landed in Bengaluru as a recent convert to the social sciences with, I now think, naïve questions about how and why consumers adopt environmental behaviors and become environmentally engaged.[49] I spent the first few months interviewing middle-class individuals and families who practiced and promoted green lifestyle practices like bicycling, purchasing and growing organic food, and home composting.[50] Once I realized that these practices were embedded in community formations such as neighborhood networks, online email groups, Facebook groups, and citywide clubs, I directed my attention to learning more about these communities of practice. I complemented qualitative interviews with visits to people's homes and to neighborhood recycling initiatives, observations at various public events organized around eco-friendly living, and online ethnography.

Bengaluru's metabolic story transformed my research as well. In August 2012, when the city was plunged into another garbage crisis, those women I was studying for their household and community practices were appearing on TV panels. These green lifestyle practitioners were capitalizing on opportunities presented by the city's latest garbage crisis. I followed the story and needed to acquire a new theoretical tool kit to make sense of it.[51] In addition to interviewing practitioners, I contacted representatives from NGOs and advocacy groups, waste management service providers, and informal economy member-based organizations. I engaged the Clean and Green Forum (TCGF), a citywide waste management advocacy network and public interest group. During my engagement with TCGF, I attended many of their weekly meetings (which are open to observers), went to High Court hearings with members from the forum, and participated in some of their waste management awareness drives. I also embedded myself with a neighborhood-based zero-waste initiative in northeast Bengaluru called the Care Collective. I participated in their weekly planning meetings and accompanied the group when they conducted waste management awareness sessions in apartment complexes and neighborhood associations.

Only in late 2012 did I begin to think more seriously about the question of labor and work in these emerging zero-waste initiatives. At one of the court hearings, I heard a Parisara Tanda staff member talk about waste pickers and the need to include them into emerging decentralized infrastructure. Soon, I connected with the organization and offered to help. I was invited to join them as a consultant. I started working with Parisara Tanda, helping develop website content, writing grant applications, and developing outreach materials. I spent an average of one day per week in their office for four months in early 2013, engaging with staff and waste pickers. I attended monthly member meetings and other events. I watched as they built a strategic coalition with the middle-class TCGF. Parisara Tanda and TCGF convinced the Karnataka High Court bench that including waste pickers in formal waste management was an economically sensible and expedient move: "They are the best at cleaning the city. They can do it for no money. They know how to earn from the waste itself." This statement, made by TCGF's lawyer, while casteist and classist, sought to gain recognition for the labor of waste pickers within existing regimes of valuation in the city.

After leaving "the field," I made return visits to Bengaluru in 2016, 2018, 2019, and again in 2022 after COVID-19. I conducted follow-up interviews with middle-class waste activists and Parisara Tanda staff. I visited some of the new waste processing sites that had been set up in the past few years. I also assisted Parisara Tanda in various research tasks, reviewing reports and policy briefs for the organization. My relationship with the organization continues to date, and I consider myself their ally. At the same time, as you will see, I raise some critical questions and points for reflection, of which they are aware. However, my proximity and connection to the organization is certainly something the reader should keep in mind while evaluating my analysis. I should also be clear that this is not a study of the informal recycling sector, waste pickers, or waste picking. I analyze the political claims, choices, and public discourses of a nongovernmental organization and waste pickers affiliated with it in relation to predominant urban sustainability policies and practices. I made an intentional decision not to make waste pickers the target of my study. This is an act of studying-up—prompted also by my privileged positionality, which I explain below. When I share the voices of individual waste pickers, these are from public media sources. This was a deliberate choice to amplify the advocacy and storytelling that waste pickers were already engaged in themselves in public fora.

My analysis of the class, caste, and gender politics of this story is also inseparable from my own positionality as a woman born into a Brahmin, middle-class family in Chennai. Throughout the book, I reflect on my own positionality, the ways in which my biases and prejudices have been challenged by my engagement with diverse actors, and my struggles to reconcile what I witnessed with what I thought I knew. My research and engagement with and alongside waste pickers politicized me, finally forcing me to confront the ways in which I had been socialized to ignore, sanction, and practice casteism in my own life. Since 2013, I have been struggling to confront the Brahminism in my own body. I started eating meat and reading the works of key Dalit scholars and activists in an attempt to unlearn caste and engage my family about our collective complicity in perpetuating casteism. This will be a lifelong journey. My work as an educator of justice, community, and leadership in Northern California has also helped me think more deeply about questions of racialized oppression in environmentalism and sustainability, and about research justice. I offer this self-examination to be transparent about the conditions under which I make these claims.

1 Consuming the Clean and Green City

I meet Swamy in his eco-home in one of Bengaluru's most expensive central neighborhoods. Trained as an engineer and today the head of a manufacturing firm, Swamy describes himself on his website as a "foodie, animal lover, and environment nut." Swamy calls his home Kachra Mane. The phrase loosely translates to "waste home," referencing the incorporation of materials once discarded as waste into its construction. A beautiful space, Kachra Mane spoke to my own aesthetic sensibilities. Large airy windows were shaded by a canopy of old-growth trees. The inside was sparsely but tastefully furnished.

As I took a seat by a window on a handloomed cushion, Swamy sat opposite me and told me that his home had recently been featured in an interior design magazine for its low-carbon, low-cost construction. He beamed with pride as he described the two toilets that had been outfitted with secondhand commodes obtained from a shop that reclaimed materials from demolished buildings. The roof was built with corrugated bamboo, the floor was simply polished cement, and most of the walls were windows. As we chatted, a monkey peered into the house. It was an idyllic setting, a wonderful respite for this tired ethnographer who had just traversed Bengaluru's hectic roads to interview Swamy about his green consumption practices.

Swamy sees himself as an avowed and dedicated environmentalist. His website offers tips on how to reduce energy and water use in the home, including a particularly descriptive entry on how to reduce ironing (do not iron above the crotch). He is an engaging speaker with a sense of humor. Schools, colleges, and corporate offices invite him to give talks on how to live an eco-friendly life in Bengaluru. His insistence on wearing what he calls his cyclist uniform—a bicycle helmet and bike shorts—in these professional and civil society spaces makes him conspicuous. Indeed, one of the first things Swamy tells you when you meet him is that a bicycle is his primary mode of transport. He also occasionally uses autos and buses, but he almost never drives a car. He knows that this makes him different, maybe even strange, in a society where the car is a symbol of respectability and propertied citizenship (Baviskar 2011).

I asked Swamy about how he came to change his everyday consumption practices:

> I was cycling until about high school. And like everyone else, I quit. After that I became a standard urban climber. So, when I started working, when I could afford to, I bought a moped, and then I could afford more and I bought a scooter, and then . . . and then a Ford Fusion, and with that I stopped. . . . I started cycling about twelve years ago. When I hit forty, I hit sense and realized it doesn't make sense to add to the condition.

Swamy is not alone in his eco-lifestyle choices. In the past decade, consumer-based environmentalism has taken root in Bengaluru. In a departure from slum removal and urban ordering campaigns that targeted the city "out there," it appears today that some sectors of Bengaluru's new middle classes[1] are problematizing their household consumption practices in relation to the deteriorating "condition" of urban and global environments. Bengaluru has become an incubator for green consumption practices, with vibrant communities emerging around bicycling, organic food consumption, urban gardening, and zero-waste management.

Embodied experiences of local environmental change, as well as circulating discourses of global ecological crises, motivate these eco-practitioners. Bengalureans are growing food on rooftops and balconies to avoid the pesticides and chemical fertilizers in commercially grown food (Frazier 2018), and as an act of resilience in an unstable urban ecology (van Holstein 2019). Composting and recycling have emerged as ways to take back personal control over Bengaluru's garbage woes (M. Anantharaman 2014). Cyclists like Swamy point to health, fitness, and concern about climate change as motivating factors, while also articulating pleasure at the speed and flexibility they experience traversing Bengaluru's traffic-choked roads on a bicycle (M. Anantharaman 2017). Families avoid prepackaged meals or limit food waste because it has always been in their traditions to eat fresh food every day (Ganguly 2017). Across the board, they recognize the impacts of their consumption choices and want to do their bit to "save the planet" while improving their city.[2] They are not alone. Worldwide, members of the middle classes and the elite appear to be trying to offset the environmental damage produced by their consumption by attempting to modify their everyday practices to become more "sustainable."

At first glance, these lifestyle-based environmental movements appear to be a radical departure from the escalating consumerism characteristic of India's new middle classes, but closer examination reveals that green consumption is also constrained and driven by social status concerns within a highly unequal and rigid class/caste order. I repurpose the term *performative environmentalism* to understand the cultural politics of everyday environmentalism. Drawing on Erving Goffman's theories of performance, sociologist Jessica Gullion (2015) coined the term to describe how fracking activists in Texas put emotions, fear, and suffering on public display to demonstrate the pain of living in a polluted region and thus gain recognition for their environmental claims. More colloquially, some media and cultural commentators have also used the phrase to emphasize the limits of small, individualized acts of behavior change to tackle big structural problems like climate change (Smith 2020). This more colloquial usage derives from the pejorative term performative activism, which dismisses social media slacktivism as ineffective and primarily about self-aggrandizement.

My encounters in Bengaluru have convinced me that we need another way of thinking about performative environmentalism. Performative environmentalism offers a way for ordinary people and nonexperts to establish ecological legitimacy, which is the moral authority to claim that one is doing right by the environment and can speak for the environment.[3] Individuals, acting collectively, perform environmental practices to demonstrate themselves as legitimate environmental stewards. They usually perform environmental practices because they are motivated by environmental concerns. But the manner in which they validate their own and each other's' environmentalism is modulated by status pressures operating within a rigid class order. Performative environmentalism should not be understood as a feint to maintain status by those engaging in it, or as a ruse to cloak class-based priorities as issues of the common good. It is sincere, driven by genuine feelings, and ideologically informed. It *nevertheless* reproduces inequities because what precisely becomes recognized and validated as an environmental practice is contingent on the class/caste position of the person performing the practice, what the practice looks like, and how it is discussed.

Cultural sociologist Pierre Bourdieu's (1984, 2002) concepts of cultural capital, habitus, and distinction offer a useful framework for exploring the relational poverty politics of performative environmentalism. When everyday so-called green consumption practices as performed by high-cultural-capital consumers become the dominant means of establishing ecological legitimacy in the city, working-class groups who do not embody the same "feel for the game" are denied ecological legitimacy *even if* they do the very same practices in their everyday lives. Green lifestyle communities in Bengaluru perpetuate an antipoor sustainability through class-based distinction practices that portray only well-to-do sustainability practitioners as ecologically legitimate, othering the poor and deepening stigmas over poverty. Thus, even when the middle classes problematize environmental crises as emerging from unchecked consumerism and wish to change their behaviors to ameliorate these issues, their sociocultural biases and status anxieties render their environmentalism exclusive, casteist, antipoor, and, ultimately, ineffective at challenging the very economic orders they critique.

Repeated calls for sustainable consumption the world over have not led to a shift away from consumerist lifestyles. Rather, new forms of greenwashing are the norm. Political scientist Peter Dauvergne (2016) calls this the *environmentalism of the rich*, emerging from the traditions of moderate Western environmentalism and reflecting commodifying, co-opting, and individualizing effects of neoliberal capitalism on radical environmental movements. My articulation of performative environmentalism helps us understand why environmentalism of the rich has become the dominant expression of environmentalism in the twenty-first century, and consequently why several decades of local activism have failed to deliver absolute reductions in resource consumption and greenhouse gas emissions, *even when people have good intentions*. This failure is due to the aesthetic and performative biases and exclusions of sustainable communities, historically grounded and reinforced as they are by everyday performances that grant ecological legitimacy unevenly. By being accessible only to high-cultural-capital groups, these performances marginalize the voices of those most affected by the status quo of consumer capitalism and apply a green veneer to problems of overconsumption and environmental injustice.

Individualized Sustainability and the Question of Consumption

Purchasing eco-certified products, using reusable shopping bags, installing solar panels, commuting on bicycles, and other green lifestyle practices are now integral to the repertoire of the eco-aware in global cities. Even the mainstream United Nations Sustainable Development Goals now emphasize sustainable consumption. The fields of behavioral economics, social psychology, and green marketing have developed a sophisticated literature on the conditions under which people buy green products and modify their energy, food, and mobility choices. Corporate, community, and state-led initiatives aimed at modifying consumption behaviors through education and behavioral nudges are increasingly common, particularly in Europe.[4]

While the individualization of environmental responsibility is problematic because it shifts the focus away from institutions and policy, there is also a growing consensus that everyday consumption is a key domain for sustainability action.[5] People of high socioeconomic status have disproportionate influence in locking in energy-driven greenhouse gas emissions through their household consumption and their investments (Nielsen et al. 2021). Growth in affluence has increased resource use and emissions faster than can be mitigated by technology (Wiedmann et al. 2020). The message is clear: the global "polluter elite" need to cut back on their lifestyle emissions (Kenner 2019). Despite this, governments and businesses continue to promote the expansion of consumption to keep GDP growing. Robust consumer spending is key to economic growth and stability in capitalist economies, a reality that sits uncomfortably alongside the moral implications of affluence-driven climate change. Extensive waste/wasting is also essential to growth-focused capitalist economies because today's waste serves as tomorrow's profit frontier.[6]

Escalating consumerism in places like Bengaluru is a global and local environmental issue. A 2015 report published by the Global Inequality Lab shows that emerging countries make up one-third of the top 10 percent of greenhouse gas emitters in the world.[7] Criticism from the Global North about overconsumption in the rest of the world has the stench of hypocrisy as rich countries still account for most of the worlds' carbon emissions. Yet, from 2000 to 2015, the carbon emissions of the elite and middle classes of emerging countries grew more than those of any other group, reflecting their increased incomes and access to global markets and consumer goods.

Growth-boosters like McKinsey celebrate these growth trajectories, arguing that the globalization of Western lifestyles is good for economic stability and a buffer against the slowdown of consumer spending in the Global North (Ablett et al. 2007).

My family is a case in point. My father was a civil servant, my mother a journalist. My parent's occupations, caste, education, and income of 4,000 rupees a month placed us firmly within the middle classes in 1986, when I was born. Back then, we lived in a 700-square-foot government flat. The only household appliances of note were a mini-refrigerator and our *mixie* (blender), both wedding gifts. My mother and father eventually bought mopeds—two-stroke motorbikes popular in Chennai. Sometimes, the four of us would pile onto my father's moped to go visit relatives in other parts of town. I would stand in the front, ensconced between my father's legs, while my mother rode pillion, holding my infant brother in her arms.

We acquired our first family car in 1992, a secondhand Fiat 500 bought for 45,000 rupees, after taking a loan from a family member. Ten years later, we upgraded to a new Maruti 500, taking advantage of the expanded availability of credit in liberalizing India. In 2001, we bought a washing machine, one that I would wake up at 6 a.m. to use, as that was the only time we had running water during the summer months. In 2002, we bought a PC for my mother to use for her new job as an editor of an online newspaper. In 2005, my parents installed our first air-conditioning unit.

In 2022, my parents still live in a two-bedroom flat, but it is replete with not one, not two, but five air conditioners, a large refrigerator, a washing machine, a microwave, a coffeemaker, two air purifiers, a laptop, two tablets, and several other small gadgets. Global brands like LG, Samsung, and Whirlpool are well represented in their home, a far cry from the Indian-made Godrej fridge of my early childhood. Year after year, my father's income kept rising as the Indian government adjusted pay scales. The online news revolution boosted my mother's income. The fruits of their labor and luck? Five air conditioners working tirelessly in the summer months to keep them cool and comfortable. And while their footprint remains modest in comparison to my own living in a centrally heated apartment in California, it certainly towers over that of their household help, who live in a nearby slum.

India's economic growth, defended vigorously at every global climate summit as crucial to alleviating poverty, has benefited dominant-caste, middle-class, urban Indians more than any other social group (Oxfam International

2019). My educated and securely employed family's fortunes have improved with economic liberalization, but many poor Indians, whether rural or urban, have not seen their incomes increase. Income inequality has grown during thirty-five years of neoliberal, market-oriented development policy, producing an uneven distribution of emissions. India's emissions inequality is laid bare in the findings of a 2021 study measuring consumption-based carbon footprints across income classes: high-expenditure households emit seven times as much carbon as low-expenditure households (Lee, Taherzadeh, and Kanemoto 2021).[8] The damning conclusion is that India is "hiding behind the poor" in shirking its climate obligations, just as Global North countries deflect blame onto India and China to evade their historical climate obligations (Chakravarty and Ramana 2012; Dubash 2013).

Locally, consumerism depends on and justifies the creation of spaces and infrastructures that support elite and middle-class consumption, just as those same structures threaten the livelihoods of India's more numerous urban poor. Consumption by India's highest-emitting classes—even the "sustainable" sort—produces local environmental impacts, most evident in the form of solid waste, air, and noise pollution. The urban poor are disproportionately exposed to these local environmental costs, compounding the injustice of economic inequality. An elite consumer can ride around in an air-conditioned car or buy air purifiers to protect themselves from the air pollution produced by increased car and truck traffic, but slum dwellers do not have the same recourse. The discards of urban consumerism are carted off to peri-urban dumps, poisoning the land and water of Dalit and poor communities.

Ecological Citizenship, Eco-Privilege, and the Othering of the Poor

But people like Swamy are trying to buck this trend. For him, living in a house that is built to last only twenty-five years (he plans to let the land rewild after), eschewing private cars, and composting his waste is part of prefiguring an alternative way of living in the city beyond consumerism.[9] Swamy's endeavors are not solely focused on changing his own behaviors. Instead, it is evident from his efforts to spread the word about his low-carbon life through talks, his website and blog, and the work he does in his neighborhood promoting zero-waste practices, that Swamy is engaged in a collective project of remaking life and reimagining progress. People like Swamy who have decided to become more sustainable see themselves as citizens who are contributing to better cities and better planets. It is an expression of ecological citizenship.[10]

Bengalureans who are giving up their cars to ride bicycles and quitting well-paying jobs to grow organic food are quite different from the new middle-class consumer idealized by neoliberal India. The neoliberal middle-class consumer is not only economically productive, but also a reliable consumer of every commodity globalized India has to offer. A mutually constitutive relationship between consumption practice and class identity is key to contemporary formulations of India's middle classes as a class-in-practice, which political scientists Leela Fernandes and Patrick Heller describe "as a class defined by its politics and the everyday practices through which it reproduces its privileged position" (2006, 497).[11]

Bengaluru's elite cyclists, organic food growers, and zero-waste practitioners, like their counterparts in the United States and Europe, are overwhelmingly high-cultural-capital consumers embodying what Carfagna et al. (2014) term an emerging high-cultural-capital eco-habitus that is central to their identity projects and strategies for claiming status and distinction. Eco-habitus, an evolution of high-cultural-capital consumption repertoires, has developed in response to the scale and severity of global environmental crises.[12] The emergence of high-cultural-capital eco-habitus could be seen as a positive development for planetary health. If high-cultural-capital individuals see green consumption as an integral part of what people like them do and care about, then high-culture consumption might become less resource and carbon intensive over time. If Bengaluru's new middle classes, who other social groups seek to emulate (Ramakrishnan et al. 2020), were to make judgments using ecological criteria, use discourses of ecological impact, and deploy an ecological consciousness, sustainable ways of living might become the vogue across India. This is the promise of sustainable consumption: that it is both integral to and generative of broader social transformation.

Critics argue that the enthusiasm for green consumption is misplaced, that the rhetoric and practices of middle-class sustainability distract from more systematic critiques or radical challenges to the neoliberal economic and political order.[13] Its most ardent detractors describe green consumerism as a kind of greenwashing that sustains economic growth and extractive capitalism. Its proponents, meanwhile, have sought to redeem green lifestyle politics by pointing out that collectively coordinated consumption strategies offer a way for consumers to do politics by "voting with their dollars" or by prefiguring alternative ways of living (Schlosberg and Craven 2019). I offer a different perspective. The reason that well-intentioned sustainable consumption becomes easily co-opted by capitalism is because of its performative and aesthetic biases that favor elite and upper-middle-class participation in environmental politics over that of other social groups. The biases are sociocultural, deeply embedded in the organization of everyday life.

Performative environmentalism is a regime that inadvertently excludes the working poor from participating in environmental politics, further eroding their rights to the city. In making this claim, I am drawing on insights from relational poverty studies, which argue that poverty should be understood relationally (Elwood, Lawson, and Sheppard 2017). Poverty is not a product of individual failure, but rather it is constituted by sociocultural and political-economic structures and reproduced through everyday practices. The antipoor aspects of performative environmentalism are on stark display in Bengaluru, where such green practices as growing food, segregating waste, or commuting by bicycle are quotidian, livelihood-generating activities among India's poor and working classes. Yet these vernacular practitioners of sustainable lifestyles are rarely acknowledged as environmental actors. They are not invited to participate in green consumption communities, and certainly not to give lectures.

Deconstructing the conditions of possibility for the practice of sustainable consumption, as well as the discourses associated with these practices, helps us understand why this is the case. The language of sustainable consumption assumes access to forms of privilege that allow consumption in the first place. To reduce consumption, one has to have been, at one time, consuming at a higher rate. For many elite environmentalists, doing green practices every day provides a sense of purpose and progress without challenging their class status and power. It is an everyday salve that soothes the ecological violence of the neoliberal economic order. Poor people cannot participate in this discourse because they lack access to the privileges that fuel its emotional meaning.

In Bengaluru, the performance of green as a high-status practice is directly enabled by class privilege and furthers class distinction. Sustainable consumption practitioners legitimize their bicycling, recycling, and gardening practices by actively distancing them from the livelihood practices of the poor. In doing so, they *other* the poor, increasing the stigmas around poverty. When green consumption becomes seen as a high-cultural-capital practice, those who lack dominant cultural capital cannot perform greenness in the same way. They are consequently denied ecological legitimacy. An unquestioned embrace of performative environmentalism silences working-class and urban poor communities in urban environmental politics, the very constituencies most likely to have a critique of state and corporate capital and of neoliberal urban development policies.

In calling out the aesthetic and discursive biases that deny ecological legitimacy to working-class cyclists and waste pickers, it is not my intention to naturalize or valorize necessity-based environmentalism. Rather, it is to call out the specious nature of elite environmental claims, which become the basis of patronizing and disciplining poor people in the terrain of sustainability. It is performative environmentalism and neoliberal sustainability that ensure the naturalization and perpetuation of the environmentalism of deprivation because the performative sustainability of elites will always need the need-based sustainability of the poor to give its claims content.

Discursive Boundaries: The Ethical Elite

Practices of distinction are key to maintaining middle-class identity: because the definition of "middle class" is ambiguous and malleable, those who want to claim this status actively seek to distinguish themselves from the lower orders, both on the street and within the home. From the other direction, identifying as middle class gives the elite and securely propertied the capacity to speak for the nation and city, as the social group to be relied on to advance what is good for everyone (Baviskar and Ray 2011). These practices of distinction depend on long-standing forms of caste, religion, and linguistic differences and on new forms of consumption. Consumption practices that maintain middle-class distinction are particularly visible in Bengaluru, where sprawling malls, gated communities, and car-clogged roads symbolize the new middle-class lifestyle. Practices of distinction include shopping in malls, wearing branded clothing, employing domestic workers within homes, and travel by car or motorcycles on city roads.

Bicycling and a commitment to zero waste disrupt two prominent practices of middle-class distinction. To compensate for these disruptions, the middle class invoke environmental discourses to separate *their* bicycling, recycling, and composting practices from the quotidian practices of the poor. They simultaneously invest resources to make their bicycling and waste management practices *look* and feel quite different from those of the poor (see figure 1.1). In doing so, they create *defensive distinctions* that separate them from the poor and penurious working-class cyclist or recycler and the unethical, materialistic middle classes. They fashion an identity that is simultaneously ethical and elite, enhancing their ecological legitimacy while retaining class status.

Elite and ethical identities are constructed and maintained by deploying inherited and accumulated economic, social, and cultural capital. Following Bourdieu (1984), a person's social status in any given setting depends on how much access that person has to each of these capital types. Economic capital is defined in monetary terms, simply the resources that

Figure 1.1
Spandex cyclist and itinerant buyer. This picture captures two types of cyclists one might encounter in Bengaluru. On the left, a spandex-clad cyclist on a mountain bike on a recreational ride. On the right, a man who is carrying plastic buckets on an older bicycle. The second cyclist is likely an itinerant buyer, who goes from home to home, exchanging recycled plastic buckets for paper and glass scrap.

individuals can amass and deploy in their everyday practices. Bengaluru's ethical consumers enjoy lifestyles of comfort and convenience, enabled by high incomes obtained by jobs in IT or other well-paying sectors. Consequently, they can invest both time and money to make bicycling, urban gardening, waste management, and food consumption practices more safe, convenient, and status-affirming. For example, middle-class cycling evangelists go to great lengths to convince potential converts that becoming a cyclist would not come with loss of social status or reputation. Swamy explains:

> People like me must start cycling. In my talks, I say, I'm like you. And normally I wouldn't be so immodest. I tell people I own my company; in the industry we are the top brand in India. I tell them all that to make them understand that—this guy is high up so why can't I cycle. Sadly, it is needed. Yes, I tell people that nobody has lost respect for me. I am well off in my industry and I give talks in industry meetings. It hasn't affected my company's brands and its ability to sell products.

By emphasizing how he, a person who has "made it," has adopted cycling as his main form of commute without damage to his economic prospects, Swamy makes clear that it is the poor cyclist, not bicycling, that is low status. Once the bicycle has been separated from its association as the poor man's vehicle, it becomes appropriate for elite consumption. For middle-class cyclists, this means buying high-quality bicycles and specialized gear to make bicycle rides safer and more pleasant.

Studies of zero-waste, urban gardening, and organic food practices reveal similar dynamics. Zero-waste management practitioners, many of whom live in gated enclaves, invest in specialized composting solutions and recycling equipment to signal that their waste management practices are scientifically informed. Urban gardeners buy land in surrounding villages or have expansive terraces and balconies in their single-family homes to experiment with new ways of growing food (Frazier 2018; van Holstein 2019). Middle-class and elite consumers are investing economic resources both to make their green practice practicable and to make it look different from the necessity-oriented vernacular practices of the urban poor. In doing so, the stigmas associated with the practice recede.

Green practices like bicycling and zero waste are anchored in communities whose formation is facilitated not only in online social media spaces, but also in offline forums like neighborhood and apartment associations, workplaces, and social events. These communities are manifestations of social capital.[14] Communities serve as critical sites for social learning, where skills and knowledge are shared, encouragement is provided, and successes celebrated. Communities also serve as boundary-making spaces.

In Bengaluru, green talk functions as boundary work on several levels. First, because these communities are exclusively middle class, they express their commitment to sustainability primarily in English, rather than the Kannada, Tamil, or Hindi preferred by the city's working-class groups (see also Erler and Dittrich 2020). Second, in these communities, people learn to talk about their eco-lifestyle in certain ways that perform ecological legitimacy. This environmental talk identifies an environmentally conscious person as someone who acts ethically not because they need to, but because they want to.

The following text, taken from the website of one of Bengaluru's most popular bicycling communities, reveals this dynamic:

> It's a general notion in our country, when someone who spots a cyclist they feel He/She is cycling either for fun or they cannot afford to buy motorcycle/car but the same cyclist cycling with a Go Green-Tee can pass on a clear message that He/She is cycling for a cause. The print on the Tee is self-explanatory and doesn't require any briefing on the cause. YOU GET BACK U'R RESPECT WHILE U WEAR THIS GO-GREEN TEE & RIDE CYCLE.

The founder of this group is a first-generation Bengalurean who moved to the city from a small town to work in IT. His successful career increased his net worth substantially; at one point, he owned as many as six cars. He was a leader within his family and his community. He decided to become a cyclist to reduce his environmental footprint after seeing the movie *An Inconvenient Truth*. Many people in his life met this decision with surprise and disapproval. When I interviewed him, he told me that the T-shirt was his way of combating the intense criticism he received when he first began to bicycle. He decided to market the T-shirt and build a movement to popularize bicycling in the city. By emphasizing that bicycling, for the middle classes, is a voluntary act, adopted not just for personal benefits like fitness but also for planetary stewardship, the practice is elevated to an ethical choice. By talking about "going green by going cycling," bicyclists also distinguish themselves from car drivers, whose continued patronage of automobiles they take as evidence of their apathy to environmental problems.

Green talk carefully distinguishes between those who engage in resource-conserving behaviors out of necessity and those who come to these acts out of an ecological consciousness. This narrative, perhaps accidentally, down-plays the fact that individuals who could take up these time-consuming green practices voluntarily had already amassed wealth and resources. One of my interviews, for instance, was with a woman in her mid-thirties who had left her IT job to look for work in the environmental sector:

Manisha: So you've been interested in environmental issues for a long time.

S: Yes, for a long time. And also the reason why it took me a while to break out of IT is the money factor. I did want that financial freedom. So then I realized I had enough, earned enough.

When S realized she had "earned enough" to be financially secure, she was able to downshift her life. She quit her IT job to start working for a bicycle store. Others I interviewed who had left their IT jobs to start eco-businesses or work in the nonprofit sector did so after either accumulating savings or owning property. Yet others relied on well-employed family members for financial security. Most of them had advanced training and were confident that they could find a job in the IT sector again at a push. Being able to say that one was doing these things not because one had to, but because one cared and wanted to, was key to middle-class, ethical environmentalist identity formation.

These distinctions were primarily defensive, in that they were not so much about enhancing social status within one's class group, but rather about not losing respect or face with family, friends, or colleagues. In a country where ownership of a car or motorbike is a marker of respectability and even marriageability, and engaging with waste is associated with caste stigma, taking up bicycling or recycling has social costs. Hence, it becomes all the more important to middle-class people to explain that their deci-sions are made out of choice and not necessity. They are not accidental environmentalists (Kennedy and Horne 2020).

A third form of symbolic capital, cultural capital, makes performances granting ecological legitimacy convincing.[15] Members of green consump-tion communities work as lawyers, managers, accountants, urban planners, researchers, and media professionals. Access to this expertise through kinship networks, educational networks, and neighborhood clubs helped bicycling and waste management practitioners gain more publicity for their activi-ties (journalists and influencers), generate research and data to support their schemes (accountants and researchers), draft and pursue legal cases in the courts (lawyers), and devise plans to improve bicycling and waste management infrastructure (urban planners).

Cultural capital in turn became a means of othering and distinction. Simply put, the capacity to talk of these practices as *environmental* practices, and to articulate their importance to planetary well-being, personal health, and the city was, in and of itself, a product of education. Zero-waste management practitioners drew on discourses, meanings, and symbols of global environmental protection, caring for city spaces, public interest, and citizenship to elevate their waste management practices as ethical, thereby distinguishing them from the types of recycling practiced by members of the urban poor. They also used technical language and scientific concepts to stabilize their practices as suitably modern and appropriate for a self-identified high-tech city. For example, new middle-class waste management practitioners were often found discussing the relative merits of aerobic versus anaerobic composting.

Similarly, a study of urban gardeners in Bengaluru found that middle-class gardeners position themselves as carriers of knowledge, sophistication, and new technologies, and in turn describe their hired gardeners as backward (van Holstein 2019). Non-middle-class members of the Vanniyakula Kshatriya caste, who far outnumber middle-class organic gardeners, carry out most of the urban farming in the city. However, they are not recognized as "good" urban agriculturalists because of their cost-necessitated use of pesticides and fertilizers (Frazier 2018). Another study looking at millet consumption and organic food shops argues that the resignification of millet as a lifestyle superfood, enacted through new ways of talking about its health benefits, created and maintained symbolic boundaries that excluded lower-middle-class people (Erler, Keck, and Dittrich 2020). This restyling of long-standing provisioning and consumption practices, such as the eating of millet, a food indigenous to the drought-prone Deccan plateau where Bengaluru is located, disassociates the practice from its rural and nondominant-caste origins. Although the motivations and anxieties that propel individuals to these forms of green consumption might be diverse, the outcomes are consistent: they reify an existing social order.

A global order that sees rising private consumption as a more important indicator of success than public wealth in turn encourages performative environmentalism. Elite claims to environmentalist status are backed up by corporate and capitalist agendas that welcome environmental campaigns that leave political economy or the interests of capital unquestioned. In US cities, municipal governments promote community gardens, farmers markets, and bicycle lanes because they represent a form of "sustainability capital" that makes a neighborhood more attractive to wealthy, white residents, and therefore to real estate developers (McClintock 2018). This is what sociologist Miriam Greenberg (2013) calls "market-oriented" sustainability, a strategy to protect the conditions of capital accumulation in an ecologically unstable world by turning sustainability into a green-growth strategy. When cities develop bike and public transit infrastructure, the wealthy gentrifiers who move into these neighborhoods actually emit more carbon than those they displace, simply because of their higher consuming capacity (Rice et al. 2020). But regardless of their actual carbon footprints, it is the cycling techies and organic moms who are seen as sustainable, rather than the immigrants living five to a room, driving gas guzzlers to work from their suburban homes.

Aesthetic Boundaries: The World-Class City and the Garden City

Performative environmentalism maintains class structures by drawing aesthetic as well as discursive boundaries.[16] These judgments of aesthetic value are not imposed on people per se, but rather on the spaces they inhabit, a form of territorial stigmatization (Wacquant, Slater, and Pereira 2014).

What types of spatial imaginaries animate these aesthetic judgments, and where do they come from? Elite visions of what the city should look like oscillate between preserving the green heritage of Bengaluru's past and emulating world-class cities like Singapore. The "world-class city" is a dominant spatial imaginary among Bengaluru's corporate classes, animating their environmental politics. Marked by such motifs as glass-fronted towers, paved walkways, and a high-tech metro system, the world-class aesthetic is most visible in the south and southeastern parts of the city that house tech companies and their workers. Its proponents, Bengaluru's IT-entrepreneur class, whom Govind Gopakumar (2020) calls "Brand Bengaluru," see livability as connected to modernity. Relatively new entrants to the city, these technocrats see Bengaluru's redemption as dependent on the city's expedient adoption of new ideas flowing through the global circuits of capital and people. Rendering the city world-class involves investments in urban infrastructure as well as discarding things and people that make the city look shabby, disorderly, or backward.

The aesthetics of the world-class city criminalize the activities of the poor when they encroach on public lands and waters, while ignoring elite illegalities such as tax evasion or building in ecologically sensitive areas (Ghertner 2015). The former look out of place, while the latter fit into the broader vision of what a "world-class city" is supposed to look, smell, and feel like (Truelove and Mawdsley 2011). The state, moreover, is eager to restrict environmental action to slum removal, cleaning up or preserving some green spaces, as these actions do not threaten its neoliberal growth agenda. The courts and urban planners use codes of appearance to permit environmentally destructive development that looks good and enhances the appearance of the world-class city at the expense of lakebeds, marshlands, and the slums that occasionally occupy these spaces.[17] Instead of authorizing development on the basis of compliance with existing regulations, planners take their cues from loud civil society groups, environmentalists, and the wish images of global urbanism to raze informal neighborhoods.

The eco-habitus of people like Swamy stands in tension with this imaginary of a world-class city. Instead, many members of these green lifestyle communities problematize the impacts that IT-led development has wrought on the city. Swamy continues:

> Bangalore used to be a hub of art, Indie culture, and high technology in terms of defense. In a way, it had all the qualities of being a knowledge city. Something like Boston. But what happened is that there was an article in *New York Times* that called Bangalore Silicon Valley. So Bangalore is now called India's Silicon Valley because of that. It is essentially IT. IT sucks, it sucks up all the creativity. Good musicians or artists or poets or physicists are sucked into this industry. It's the money. And Bangalore essentially has become a huge IT hourly wage city. It is all making software products. . . . So this city has become all about what the IT industry needs, and about the infrastructure that is required to sustain IT. That's what the city is. There used to be a culture . . . but now there is a shift to a more materialistic mindset. We are basically measuring people by the size of their cars. . . . Physically there's a huge explosion of traffic. Sheer number of vehicles killing the facilities pedestrians need. To be a pedestrian, you need trees. To be a cyclist, you need trees. But cars don't need trees. They hit trees.

Swamy, like several other IT and ex-IT people I interviewed, expresses regret at how IT has changed the city, even though his own career has benefited from the IT boom.[18] While IT and other privileged workers acquire the success symbols of development and modernization, such as private automobiles and apartments in gated complexes, Bengaluru's crumbling infrastructure groans under the weight of their aspiration. In turn, people who had already "made it," like Swamy, sought to disentangle themselves from the consequences of their consumerism by advocating a low-carbon life. Instead, they want to revitalize what was once special about Bengaluru through their work, re-localizing their environmental imaginaries. For eco-oriented Bengalureans like Swamy, the dominant desired eco-aesthetic is that of the garden city.[19]

Comprising images of lush parks, lakes, tree-lined avenues, and walkers listening to the chirp of birds and breathing cool air during their evening strolls, this garden city imaginary sees Bengaluru as a city with and in nature, filled with sensible, educated, genteel people (Nagendra 2016). Indeed, one can still experience lush greenery in Bengaluru's colonial core, in Cubbon Park, Lalbagh gardens, and in neighborhoods like Malleswaram and Basavangudi, where major roads are canopied by large flowering trees (Nagendra and Gopal 2010). IT-led development has denuded the city's greenery, with trees felled for road-widening and metro-construction projects, angering environmentalists like Swamy. The older colonial and postcolonial middle classes of Bengaluru have long seen Bengaluru's verdant greenery as its distinguishing factor, making the city more salubrious than most other Indian metropolises. But, like the world-class city, the garden city imaginary indulges in its own forms of aesthetic criminalization.

In her history of Bengaluru's journey into metropolitan status, historian Janaki Nair (2005) highlights how this longing for a garden city draws on nostalgia for a time in which Bangalore was a "placid and restrained" paradise of wide, tree-lined streets and well-planned neighborhoods filled with respectable middle-class folk. Never mind that Bengaluru's past as a middle-class garden city is as Nair puts it, a "mythicized past" rather than a material reality, a product of forgetting. The city has long been a site of trade and industrial production, home to a diverse group of nondominant-caste, non-Hindu populations. This imagination of and desire for a green past, present, and future is hardly limited to Bengaluru. Rather, as geographer Hilary Angelo (2021) convincingly demonstrates, parks and urban greening have been part of bourgeois subject formation in multiple places and time periods. "Green as good" is now in the very DNA of contemporary environmentalism.

In Bengaluru, middle-class eco-habitus is both constitutive of and constituted by appeals to the garden city. In postcolonial India, middle class-ness is expressed through practices that distinguish this class group from the masses through associations with modernity, self-restraint, order, and rationality. To be middle class is to inhabit a place that is free from the pollutants that would mar this spatial and ideological order, whether those pollutants take the form of industrial activities, the poor, Dalits, or other oppressed castes. Today, new middle-class Bengalureans secure themselves in gated communities or in older neighborhoods like Malleswaram or Basavangudi with tree-lined streets, partially shielding themselves from Bengaluru's many industrial parks, commercial areas, and slums. Entry into these enclaves is controlled; maids enter through back gates. Watchmen keep undesirables out. The "help" have separate toilet facilities to maintain caste boundaries.

In these private spaces, the spatial imaginary of the garden city is partially realized, just as Bengaluru's parks and tree-lined avenues are sacrificed to widen roads and build transit systems. In a true perversion of ecological priorities, the "clean and green" aesthetic in posh neighborhoods like Koramangala, Sadashiv Nagar, or Malleswaram is maintained by dumping the city's trash in ecologically sensitive lakebeds, forests, and in *gommala* (public common) lands that abut peri-urban, oppressed-caste villages like Mavallipura, Mandur, and Bagalur. This enclave mentality of middle-class environmentalists can only be understood through an examination of caste in the city.

Caste, Cleanliness, and Environmentalism

"Clean and green" is a common refrain in Indian cities, plastered on school boundary walls, street medians, and other public spaces. Cleanup campaigns conflate clean cities with sustainable or ecologically healthy cities. This discursive confusion fortifies the antipoor nature of middle-class environmental campaigns. Perceptions of cleanliness—and its converse, dirtiness—have long defined moral and spatial boundaries between social groups (Douglas 2003; Sibley 1995; Zimring 2017). Garbage and pollution, as well as the working-class livelihoods that are associated with them, are framed as threats to order and salubriousness (Baviskar 2019b; Ghertner 2012). Enacting this so-called environmental agenda then demands the removal of people and things who do not conform to the dominant social group's notions of cleanliness, order, or beauty. From neighborhood-level cleanup and beautification campaigns to public interest litigations that seek to remove informal settlements, cleaning up and beautifying urban space is a key mode of realizing elite environmental imaginaries (Dürr and Jaffe 2010).

Understanding how "clean as green" aesthetics stand in for ecological protection requires a brief review of the history and dominant ideologies that underpin modern environmentalism in the West, as well as the ways that religion, caste, and environmentalism intersect in Indian cities. Dominant ideas of colonial environmentalisms have long equated ecological health with "natural beauty" and painted the "other" (the nonwhite, poor, savage, and so on) as the enemy of nature. This discourse positions the enlightened privileged as defenders of nature (Guha 1989; Merchant 2003). The writing and paintings of early preservationists, wilderness enthusiasts, and landscape artists was replete with connections between cleanliness, emptiness, and nature. This ideology constructs nature as beautiful yet fragile, too easily marred by people and their activities. This association of nature with emptiness and purity is one way that modern Western environmentalism was marked by the pervasive logics of white supremacy and imperialism. American environmentalism's racist roots have in turn shaped global thinking about conservation (Kashwan 2020). Taking this broader swath of environmental history into account, we can see that aesthetics have long stood in as shorthand for what is environmentally desirable or sound.

This vein of environmentalism is also marked by forgetting and erasure. Such forgetting is key to constructing "desirable pasts" in environmental discourse, whether in the form of a Bengaluru as a mythicized garden city or an emerging eco-traditionalist discourse that paints Hinduism as inherently eco-friendly.[20] Eco-restoration projects in Indian cities seek to restore reservoirs and waterways back to "pristine pasts" and "original boundaries," ignoring how socio-technical and natural processes combine to create urban socio-natures (Coelho 2020). As Amrita Sen and colleagues find through a study of lake restoration projects in Bengaluru, "What passes as ecological restoration projects in official documents is fundamentally oriented towards the creation of enclosed, upper-class spaces, overriding concerns over collective justice" (Sen, Unnikrishnan, and Nagendra 2021, 121). Aesthetic concerns tend to supersede the livelihood needs of local communities eking out a living at the waters' edge and sometimes cloak more vicious prejudices.

Cleanliness is also a marker of modernity. Nationalist and developmentalist campaigns led by the Indian state have long framed garbage as the dangerous "other" of modernity, the opposite of order, salubriousness, and quality of life.[21] The Indian state uses the problem of waste to activate nationalist pride and contour claims to legitimate political citizenship. When Prime

Minister Modi grabbed a broom to metaphorically clean up India while launching the BJP government's flagship Swachh Bharat "Clean India" Mission in 2014, he suggested that "everyone" could do their bit to keep Indian cities clean and welcoming. Being clean and keeping clean in turn constructs notions of hygienic citizenship, where the model citizen is one who keeps themselves and their environment clean and orderly. Anyone caught unclean or sullying the environment (which is usually the poor, as they are more likely to lack access to sanitation services) is to be disciplined into compliance (Doron 2016; Luthra 2018).

Cleaning up or protecting nature is not just about public space or wilderness, but also often about controlling people. Implementing "clean and green" politics can mean *discarding* populations as a technique of power (Liboiron and Lepawsky 2022, 62). The fact that the groups targeted by these cleanup or conservation efforts in contemporary India are often Scheduled Castes, Scheduled Tribes, or non-Hindus is no accident, as urban environmentalism is also dominated by perspectives that deny the enduring salience of caste and religion as a form of repression, oppression, and violence in India.

In India, the pervasive role of caste structuring urban ecologies compounds the work of "clean as green." The middle classes often absolve themselves of culpability for urban environmental degradation, instead displacing blame for pollution and garbage onto the urban poor and the diverse industrial sectors that employ them (Baviskar 2011). Caste is the indigenous mechanism of oppression that fortifies this move (Reddy 2021). Stigmatized castes are easy to blame because occupational caste structures leave them responsible for the dirty jobs, such as leather tanning or scavenging. As Mukul Sharma argues in his critical intervention on Indian environmentalism, "from village to city and temple to school, caste metaphors of pollution, impurity, and dirt dominate places and spaces through imaginaries of dangers posed by the presence of Dalits" (M. Sharma 2017b, 2).

Caste legitimizes an aesthetic approach to urban environmentalism by creating a social ecology that makes malodor, dirt, and filth the companion of some communities, just as it secures "clean and green" spaces for dominant-caste cultural and economic elites (Guru 2011, Lee 2017). Anthropologist Joel Lee and his interlocutors (2017) name the spatial arrangements that disproportionately funnel waste matter and its chemical and olfactory concomitants toward Dalit communities as environmental casteism, which parallels environmental racism in the United States and elsewhere (Pulido 2017).

Environmental casteism damages the health, opportunities, and well-being of oppressed caste communities. It is also unmindful of the horrors that caste-based occupations involve, such as the risk of death posed by manually cleaning sewers, a job that kills workers by drowning them in refuse (Shankar and Swaroop 2021). Practices of untouchability also inhibit the development of public culture and common goods accessible to all people (Rodrigues 2009). Instead, environmental casteism creates a segregated spatial-sensory order that is self-perpetuating, or as Lee (2017, 488) articulates "not only do we sense our place in the caste order when we breathe, but the spatial-sensory arrangement in which we live forges us into signs that replicate the order." Compounding matters, Indian environmentalism, dominated as it is by Brahmins, has ignored caste in its analysis and instead naturalized the Hindu caste system as ecologically sound (Aiyadurai and Ingole 2021).

Elite-led cleanup campaigns often involve making *someone else* clean up the city (a point I elaborate in the following chapters). Though the middle classes might occasionally wear gloves, pick up a broom, and clean litter on a street in an expression of Gandhian commitment or consumer citizenship, these environmental campaigns critique and seek to reform existing municipal waste management systems, and particularly the people who work in them. Caste is part of these technologies of discipline, invoked to devolve toxic labor onto some, but not other, bodies, just as in other contexts, waste work is racialized and/or feminized work, carried out by refugees, migrants, religious minorities, and displaced populations (Doherty and Brown 2019). As we will see in later chapters, well-meaning efforts to clean up the root causes of garbage and pollution are belied by the often-unspoken reliance on and availability of casted and racialized labor as a dehumanizing infrastructure to externalize risk, toxicity, and pollution.

A "clean up to green up" agenda is ecologically suspect, as it mobilizes ecological arguments in a manner that isolates specific instances of waste or pollution from a broader consideration of environmental flows (Liboiron 2021; Liboiron and Lepawsky 2022). Environmental campaigns conflate cleaning up with ecological preservation and sustainability in a manner that pays scant attention to social relations or processes of spatial displacement.[22] These campaigns do not, in other words, address where waste actually ends up and or who has to deal with this waste. They rely on a double act of misrecognition: of the root causes of the environmental problem of waste and pollution, and of the lives and rights of the poor. Cleanup campaigns criminalize slum dwellers, migrant laborers, and street vendors, even as they valorize the role of citizens beautifying and greening cities through behavior change and civic action.

This useful trick of using public and environmental nuisance laws to legitimize the violent removal of people from communal spaces is part of the playbook of environmental and civic activists in the United States as well. In Seattle, homeless encampments are targeted for removal under public nuisance laws, citing accumulated trash (Rufo 2019). Camp residents defend themselves, saying that their better-resourced neighbors use their homes as dumping sites for unwanted bulky household items (Barnett 2019). In San Francisco, residents of a wealthy neighborhood sued the city under the California Environmental Quality Act to stall the commissioning of a much-needed shelter. Their lawsuit argued that building the shelter would promote "open drug and alcohol use, crime, daily emergency calls, public urination and defecation, and other nuisances." In city meetings, residents held up signs of needles and garbage, arguing that people without fixed homes were innately anti-civic and anti-environmental (Ho 2019). These campaigns operate on a simple rationale: if it looks dirty, it must be bad for the environment. And if you live near dirt, you cannot be an environmentalist.[23]

Retheorizing "Sustainable Consumption" as Performative Environmentalism

Why do cultural elites such as India's new middle classes have an easier time establishing ecological legitimacy than the poor? Repurposing the term performative environmentalism, I have argued that this is because, in neoliberal late capitalism, dominant environmental discourses associate sustainability with specific aesthetic and discursive markers constituted by the preferences and possibilities of groups with intersecting forms of race, class, and caste privilege. The everyday practices most associated with green credentials involve spending time outdoors, keeping your surroundings clean and litter-free, recycling, bicycling, and buying organic food or carbon offsets. Performative environmentalism is thus only accessible to those who have either the surplus income to buy green products or the surplus time to take on laborious environmental practices. Power confers green.

Distinction and othering become a part of the performance of green behaviors because our economic order valorizes consuming power and stigmatizes poverty. Green consumers' approach to investing in sustainability contributes to the othering of the poor and further stigmatizes poverty. Their actions are not always intended to exclude, but their defensive distinctions are key to stabilizing sustainability practices that are associated with poverty and precarity within the existing economic order. In doing so, performative environmentalism reinscribes class, caste, and race divisions. In this way, and despite good intentions, performative environmentalism becomes an alibi for global capital, perfectly compatible with a world that sees compound economic growth as the prerogative for organizing social, economic, and political life.

At the same time, enacting green consumerism is a means for the new middle class and elites to acquire and leverage power. Doing green things, via ethical consumption, is a way of acquiring social status (Barendregt and Jaffe 2014). Doing green things out of preference as opposed to necessity confers power in the form of moral authority (Kennedy and Horne 2020). Moral authority enables groups to speak for the common good, giving them what geographer Laura Pulido calls ecological legitimacy. Ecological legitimacy turns into political power, enabling some groups to exert influence on political processes that affect shared environments. Citizens who have achieved ecological legitimacy can claim to speak for the collective good and influence public policy, to the exclusion and harm of the presumed "non-ecological" poor. Green confers power.

Individualizing and lifestyling environmental action systematically hides the gendered, raced, classed, and ability-based possibilities and preferences of performing sustainability. It pretends that everyone has access to these environmental practices and then stigmatizes those who do not enact them as either ill-informed or uncaring. It turns a structural problem into a moral one. If children from low-income communities of color in the United States are found eating processed fast food instead of kale chips, their caregivers are held up to scrutiny, the racist assumption being they are not taught the value of fresh, organic food at home (Anguelovski 2015). Neither the uneven distribution of grocery stores and farmers markets produced by historical and contemporary neighborhood segregation, nor the exclusionary nature of progressive food movements that center whiteness, are invoked in the casual but influential theorizing of (white) armchair environmentalists (Alkon 2012; Guthman 2008). Dominant sustainability discourses that are embedded in and reflect casteist, racist, and antipoor epistemologies also render invisible the healthy and sustainable food practices of marginalized groups. Instead, these communities are targeted by college service-learning programs for environmental education and community gardening interventions, the assumption being that the poor and people of color just do not know enough to do the right thing (Malier 2019).

In urban India, this antipoor environmental ideology dominates discussions of sustainability. It construes the livelihoods of the poor as problematic, while turning away from the overconsumption, entitlements, illegal behavior, and immoralities of the rich. This is a capitalism-friendly environmentalism that seeks the removal of cows and street vendors from roads but does not ask why malls are built on marshlands or why agricultural lands are commandeered for the construction of highways. Slum dwellers are often unable to keep their surroundings litter-free because municipal waste collection services do not serve their neighborhoods, but these structural factors are ignored in conversations that frame the poor as lacking the appropriate dispositions for good environmental behaviors. Sometimes, as when informal settlements are razed to the ground to make way for an eco-park, antipoor environmentalism turns violent.[24] A commonsense, subtle, and insidious orientation that sees elite environmental claims as more legitimate than others undergirds these more spectacular instances of green dispossession.

Performative environmentalism is the everyday expression and legitimating structure of a form of neoliberal sustainability that seeks to solve socio-ecological crises with recycling, retrofits, educational campaigns, and feel-good consumption. However, as chapter 2 contends, everyday environmentalism can also be a gateway to greater political engagement in urban ecologies. When the women of Bengaluru's ecologically conscious middle class became frustrated with the lack of infrastructure and policy support for their eco-practices, they formed community organizations and engaged in political advocacy to change the material and social contexts of their everyday lives. Does performative environmentalism leave room for political action on environmental issues? And are the social groups who are excluded from sustainable communities able to rearticulate their environmental claims in a manner that is more legible and legitimate within this performative environmental order? The subsequent chapters explore these questions.

2 Communal Sustainability

Pramila, a commanding woman in her fifties, addresses a crowd of thirty mostly women, children, and a few very vocal men:

Nowadays of course we want everything white . . . so we want Italian equipment to take care of our Indian garbage. The poor Italian equipment only knows pizza; it does not know *huli majjuge, saaru, sambhar* [local stews and rice dishes]. It has not heard of these things. . . . The foreign equipment is not able to deal with Indian garbage. So, if we can instead just segregate [waste], it will be excellent. We want you to become the leaders. . . . What we are talking about is nothing *hi-funda*, no nano-technology, no big deal, it is very simple, . . . and the fact that you are here means you want to take that extra step and go beyond your own *mane* [home]. So please come give us your [waste generation] data. . . . The BBMP [Bruhat Bengaluru Mahanagara Pallike] is actually asking for volunteers everywhere. Now if we have 198 wards, we need 10 people per ward. So, *neeyu* [you] be the first volunteer, you be the leader in your ward, and then we will actually find that there is a potential that you and I never imagined we had. Ten years ago, I never imagined I would stand in front of you and talk like this. It just happened and we have all stepped out of our own comfort zones. . . . We are all here. . . . I thank each one of you for coming out, and we look forward to working with all of you, to make Bengaluru a garden city, *ellarum sendhu* [all together]. *Applause.*

With Pramila on stage are six other representatives from Care Collective,[1] a zero-waste advocacy group based in an older, predominantly Brahmin, upper-middle-class neighborhood in northeast Bengaluru. On this September morning in 2012, we are at the Bengaluru Recycling Habba, a zero-waste and recycling festival organized by the *Bangalore Mirror* newspaper in collaboration with the Clean and Green Forum (TCGF), a consortium of nongovernmental organizations, neighborhood collectives, environmental entrepreneurs, and individual volunteers promoting zero-waste practices in the city. As Pramila hands over the reins to Chitra, a younger member of the group, the twenty-three-year-old puts up a slideshow and tells the audience that she will start her presentation with a few questions: "Raise your hand if you have done any of these things [featured in the pictures up on the slideshow]. Please be honest." She pulls up a picture of a person throwing a candy wrapper out of a moving car. "Everyone is likely to have done this, at least once, right?" No hands go up. Next, a slide shows discarded food: "If not this, then this, wasting our food?" An audience member piped up vociferously, "No! No!" Another declared, "We never waste food!" Several others nodded and vocalized in agreement. This audience was rather unwilling to admit any personal culpability for the city's garbage woes.

Chitra, seeming bemused, continues, pulling up an image of a street-side dump replete with putrid banana leaves, plastic bags, and a lonesome cow helping herself to a snack of old rice: "Really? Okay perhaps not, maybe none of you have littered or wasted food. But all of us see this every day, right outside our houses." The audience nods and nearly everyone puts up their hand now. One person even puts up both his hands in agreement. Chitra continues, "In Bangalore, we have [a] very forgiving attitude, that is how it is going to be, *solpa adjustumadhuthera* [just adjust a little], walk around it, you know. Close your nose. I mean that's how we have been living and that is how things are." An older man interrupts her as she speaks "Madam, just a minute, why all this Bangalore, Bangalore. . . . It is like this everywhere in India." Chitra cuts him off quickly saying, "Okay, yeah, but we are in Bangalore, and we would like to talk about what we are experiencing. It would not be fair if I spoke about Chennai, Mumbai, or Delhi. I have not lived there. I have grown up here, so it is the focus." She proceeds, having gotten consent from the rambunctious crowd, holding her own among an audience of older uncles and aunties: "So how do we turn this around? All of us agree that this has to change, right? That is why we are all here. It is simple, and as Captain Planet says, the power is yours."

As Chitra's talk goes on, the images on the screen become more jarring. She puts up a picture of a cow with open innards and plastic pouring out.[2] The audience gasps. Next, Chitra displays an image of a pack of menacing stray dogs. She shows several images of the Mandur landfill, describing the terrible stench of rotting garbage and dump fires that neighboring villagers live with. Her face turns foreboding as she warns the audience about dangerous leachates from the dump seeping into the soil, poisoning groundwater. She draws a connection to global climate change and links it to local weather changes: "Forget your air-conditioned garden city, now we are a hot garbage city! And we are contributing to climate change with all the waste in our landfills!" In the short span of ten minutes, she has given a comprehensive overview of Bengaluru's environmental rap sheet. She turns her pointed finger at an audience that is now quiet and subdued. She tells people to stop using bottled water, to avoid disposable cutlery, to cycle, to drive less.

Each weekend, this tight-knit group of seven to ten people takes its roadshow on environmental disaster and citizen responsibility to apartment complexes, gated communities, or eco-festivals like the Recycling Habba. There is one man in the group, a husband of one of the women. Else, this is a merry, multigenerational troupe of (mostly) women traipsing the city telling other women about the power they possess to transform their neighborhoods, communities, and cities. For several months, I joined Sheetal, Pramila, Chitra, and others on their weekend visits. Sometimes we would do two events over the weekend, traveling more than fifty kilometers back and forth to the far-flung reaches of ever-expanding Bengaluru. During these trips, we usually met and spoke to other women who were trying to initiate or maintain zero-waste programs. While the motivations for going zero waste ranged from environmental concern to wanting to comply with new Bruhat Bengaluru Mahanagara Pallike (BBMP) rules, a commonality among the thirty-four communities I visited was the tremendous amount of work demanded of the leaders and executors of zero-waste initiatives.

How do Bengaluru's environmentalists envision and enact community and to what ends? Building on chapter 1, where I articulated the contours of middle-class *performative environmentalism* in Bengaluru, I dive here into a detailed case study of Bengaluru's middle-class and women-dominated zero-waste communities. I develop the term *communal sustainability* to describe neighborhood-based interventions into the city's waste metabolisms. Communal sustainability goes beyond simplistic explanations of environmental harm that blame lack of knowledge, bad behavior, or immorality for garbage crises, instead forging collective pathways to overcome shared environmental problems. Housewives, retired men, and other unlikely suspects deploy affective and reproductive labor to change household behavior, build small-scale infrastructures, and convene collaborative systems of governance. They buck the trend, emphasizing localism, vernacular solutions, and manual engagement in place of the world-class, high-tech, or automated as key to solving environmental problems.

Examining communal sustainability through the lens of social reproduction theory, I reveal how the socially reproductive labor of middle-class women and the working poor produces zero-waste management as a form of sustainability. In its material solutions to environmental problems, communal sustainability mobilizes metabolic divisions of community that are gendered, classed, and casted. At the same time, I caution that there are limits to seeing the performance of reproductive labor solely through the lens of governmentality or the extraction of surplus labor. What is operative here is a sense of empowerment, a building of shared identity, and an enactment of politics for those engaged in this work, which cannot be reduced to narrow economism or top-down governmentalization. Rather, under some conditions, communal sustainability, with its metabolic reliance on volunteer effort and manual labor, undermines neoliberal agendas and opens new avenues for political participation by marginalized groups in urban environmental politics.

Social Reproduction and the Sustainable Community

Who will save the city and the planet? In recent years, both policymakers and academics have placed a lot of hope on community-based initiatives to drive broader sustainability transformations.[3] Disheartened by the lack of state leadership and wary of the neoliberal push to restrict environmental action to market decisions made by so-called rational consumers, community has emerged as a happy medium—the scale at which material change can be enacted and real progress witnessed. Scholars writing from Europe and North America have reflected hopefully about how communitarian values, which emphasize the common good and social bonds over individual gain, could help democratize sustainability transitions by involving ordinary people in grassroots political action (Kenis 2016; Portney 2013; Seyfang and Haxeltine 2012). Community is invoked as an important tool for the (self) governance of environmental resources and coproduction of infrastructures, providing one architecture for place-based collective action to solve common problems (McGranahan and Mitlin 2016; Ostrom 1993; Sen and Nagendra 2020).[4] Scholars promoting pluriversal alternatives to hegemonic growth-focused development pathways in the Global South identify community dynamics as key to building new socio-metabolic configurations that can achieve environmental balance and improve the quality of collective life (Barkin 2022; Paulson et al. 2020).

How well-founded is this hope in the promise of community as a route to urban sustainability transformations? A growing literature offers a more critical assessment. One common criticism is that community-led efforts are not at the appropriate scale, either materially or in their political capacities, to effectively govern complex urban environments or remake urban infrastructures (Nagendra and Ostrom 2014). Beyond evaluating environmental efficacy, social justice questions loom. Political ecologists and other critical scholars have long cautioned against romanticizing community in environmental governance.[5] Communities can be hierarchical, exclusionary, and oppressive, just as they might empower and nourish some of their members. In India, scholars have warned that "community" in many cases is a stand-in for caste or religion, and that the primary function of invoking community is to mark boundaries and establish social hierarchies. Indeed, a closer look into many celebrated cases of traditional commons management in India reveal that "commons" were maintained either through the systematic exclusion or exploitation of oppressed castes (Sen, Unnikrishnan, and Nagendra 2021; M. Sharma 2017b). The ambiguity of "community talk" can mask fundamental barriers to broad-based collective action, such as social differences and inequalities (Catney et al. 2014).

Further, the turn to community, some have argued, is also part of neoliberal rollback/rollout processes in which responsibility for achieving national aims and social goods devolves onto community organizations (Peck 2010; Taylor Aiken et al. 2017; Walker 2011). Participatory mechanisms designed to increase community participation in urban governance can sometimes worsen local hierarchies and internal differences (Rigon and Castán Broto 2021). Ultimately, neither top-down nor bottom-up community-based environmental projects can be assumed to advance inclusion and democracy or equitably represent all stakeholders (Grossmann and Creamer 2017; Taylor Aiken et al. 2017). The social and environmental justice implications of community-based sustainability efforts are contingent and contextual—and merit ethnographic examination.

To examine internal dynamics of community in finer grain, I turn to social reproduction theory. Marxist-feminist geographers introduced social reproduction theory into environmental and development studies (Fraser

2021; Katz 2001). Geographer Giovanni Di Chiro defines social reproduction as the "intersecting complex of political-economic, socio-cultural and material-environmental processes required to maintain everyday life and to sustain human cultures and communities on a daily basis and intergenerationally" (Di Chiro 2008, 281). Sometimes termed *care work*, social reproduction is indispensable to any system of social provisioning, including capitalism, in "sustaining the human beings who constitute 'labor' and forging the social bonds that enable cooperation."[6] Building on these interventions, I outline a conceptual framework for analyzing the gendered, classed, and caste-based socially reproductive work that goes into producing sustainable households, communities, and, ultimately, cities. Such a framework can help render visible the diverse forms of uncompensated and undervalued work within the home and the neighborhood that go into greening consumption practices, resource flows, and urban infrastructures, revealing the structure and architecture of community-based sustainability.

Greening everyday life involves environmental labor or "consumption work" (Wheeler and Glucksmann 2015). Women in most contexts perform a disproportionate amount of domestic labor and care work, and this extends to household environmental work (Cairns et al. 2020; Farbotko 2018; Murphy and Parry 2021). As environmentalism has become more individualized and consumption oriented, there has been a rise in voluntary and largely unrecognized domestic environmental labor in high-consuming households. As geographer Carol Farbotko argues, women working in households have become an unaccounted-for supply of labor for the greening of capitalism. The burden of trying to ameliorate overlapping ecological crises through lifestyle change often falls on women in their roles as mothers, wives, and homemakers. Advertising campaigns and environmental organizations ask women to cook differently, recycle, and reuse more for planetary sustainability and future generations (Braun and Traore 2015).

Socially reproductive work is also raced, classed, and casted. In most middle-class and elite Indian homes, women (and occasionally men) who work as domestic helpers handle household waste (Jack, Anantharaman, and Browne 2020; Luthra 2020). Gender and class intersect to produce a particular domestic configuration in Bengaluru, where a wife or mother's job mutates from doing the housework to instead supervising it. The social relations underlying domestic work are power-laden relationships of both interdependence and exploitation. Paternalistic disciplining of the "help" is a key practice in maintaining what Raka Ray and Seemin Qayuum (2009) call "cultures of servitude." The maid or the peon are frequently framed as less competent and therefore undeserving of fair remuneration, freedom, or even self-respect, and in need of discipline and tough love, as well as limited care.

Employer-employee relations are unequal power relations made even more unequal by caste oppression. Members of oppressed castes, the "racialized others" of Indian cities, disproportionately perform waste work and other critical forms of social reproduction (Hodges 2018; Reddy 2021). As we will see, the waste management systems promoted as sustainable rely on "servants" with different names who carry out waste handling functions such as segregation and transport, which perpetuates middle-class distinction and class position. Thus, it is inadequate to consider only gender and class when examining the (mal)distribution of socially reproductive work in cities. Caste and race are central to the division of environmental labor, demanding an intersectional analysis of social reproduction (Bhattacharya 2017; Pulido 2015).

Intersectional theories of social reproduction help politicize environmental labor in relation to the broader questions of the maintenance of urban environments and the conditions of capitalist production. The Indian state and the Bengaluru municipality pursue "clean and green" cities as a strategy to sustain economic vitality and for nation-branding. However, "clean and green" is not a predestined state, but rather an imaginary that is to be realized through work. Governmental campaigns such as the Swachh Bharat Mission task housewives, retirees, and maids with producing the sustainable city through reproductive work, whether that be sorting and warehousing waste, cajoling neighbors, coordinating projects, or conducting educational campaigns.[7] In Bengaluru, zero-waste management, understood here both as a symbol of sustainability capital (McClintock 2018) that adds a "clean and green" sheen to the city and as a material arrangement that keeps waste off streets, sustains the city economically and socially. In the discussion that follows, I show how the social and material infrastructure of sustainability is produced through gendered, classed, and casted socially reproductive work, anchored in communities of practice. Infrastructural labor in turn creates opportunities for marginalized constituencies to engage in urban environmental politics.

The Making of the Zero-Waste Zealot

Why does Bengaluru remain the hot spot for community-driven sustainability in India? Bengaluru's cosmopolitan and educated citizenry have long occupied themselves with schemes to improve the city.[8] The city's unusually high level of civic engagement is connected to its history, evolution, and demographics (H. Gupta 2018; Upadhya 2017). Newly independent India selected what was then called Bangalore as the site for the country's flagship Indian Institute of Science, as well as several "high-tech" public-sector enterprises. Jobs and educational opportunities attracted a significant population of professionals with cultural capital to the city, including scientists, technicians, and educators from across the country. Many of these people settled in the planned neighborhoods of Bengaluru, spaces that were themselves a product of colonial and postcolonial modernist urban development efforts.[9] In the twenty-first century, this latent interest in realizing a modern, sanitary, and civic city has both combined and come into tension with corporate and state-led aspirations to make Bengaluru "world-class." Garbage is often the flashpoint for these negotiations.

Cleaning up Bengaluru emerged as a central governmental priority in the late 1990s, when the city was aspiring to join the ranks of global technology hubs. In a desire to attract more foreign investment to its capital city, the Karnataka state government, superseding elected municipal representatives, empowered a group of corporate elites to direct urban planning and development (A. Ghosh 2005; Gopakumar 2020). Bengaluru, from colonial times to the neoliberal era, has turned to technocrats and dominant-caste men to solve urban problems. The latest avatar of technocrat-savior is in the form of Bengaluru's ascendant IT-entrepreneur class.[10] Through extraconstitutional planning bodies, investor forums, and industry roundtables, members of this class utilize dominant cultural and social capital to gain access to state bureaucracies and planning processes. Drawing on technical competencies and the discourse of citizenship, they present themselves as neutral arbiters of what is best for the city. The Bangalore Agenda Task Force (BATF) was one such male, technocratic, and corporate-dominated extraconstitutional advisory body set up by then Chief Minister SM Krishna and headed by Infosys CEO Nandan Nilekani.

In 1998, BATF conducted a survey of "all Bangaloreans" on civic issues. Garbage was among the top public concerns voiced in survey responses, with people across the city complaining about "blackspots"—informal dumping spots for garbage that block sidewalks and crowd streets. These are visible markers of Bengaluru's failure to ascend the global ranks of dynamic, investment-worthy cities. Emboldened by "public opinion," one of BATF's first and most wide-ranging moves was an overhaul of the city's municipal waste management system (described in more detail in chapter 3). It started by eliminating roadside bins in favor of door-to-door collection by private contractors using auto-tippers and other "modern" waste collection vehicles. BATF hoped that by emulating US cities in setting up curbside collection, Bengaluru could clean up while continuing to grow its ranks of consumers and, by association, its discards. This tried-and-true strategy maintains the status-quo of disposability in modern economies, as Max Liboiron (2013) observes.

Enacting "soft-neoliberalism," BATF also sought to increase civic participation in urban governance (by which they meant exclusively middle-class participation). It convened a brigade of "concerned citizens" who could monitor the cleaning up of their neighborhoods and report any wrongdoing to the local municipal authorities. These *shuchi mitras* (friends of cleanliness) were largely women recruited from residents' welfare associations. Through BATF, the state recruited middle-class women and retirees as foot soldiers of the municipal apparatus, authorizing them to discipline sanitation workers and garbage contractors.[11]

By the time BATF was disbanded by the next chief minister, it had left lasting damage to the city's already beleaguered waste infrastructures. By privatizing waste collection in more areas, the scheme increased the power of garbage contractors, whose convivial relationships with the city's political class meant they could not easily be held accountable by ordinary citizens for failing to provide services. Service provision worsened because of uneven coverage. Door-to-door collection was never initiated in many peripheral neighborhoods and informal settlements. Even though this entire scheme was designed to eliminate the so-called blackspots, these came back in full force in many places after a few months. This scheme also did little to change how the city disposed of its waste: Bengaluru continued to transport its municipal solid waste outside the city, where it was dumped with little or no processing. Bengaluru may have become "bin-less," but urban filth and ecological damage was still ubiquitous.

While the corporate elite retreated to their large, independent homes in the city's most exclusive localities like Koramangala and Sadashiv Nagar, the ordinary *shuchi mitras*, many of them housewives and retirees, still had to contend with the garbage that assailed their senses when they stepped out of their homes. Their children still had to walk onto busy roads to avoid a dump on the sidewalk. It was from the failures of prior eras of waste management that the zero-waste management practitioner, usually a woman, emerged. There is a long history of the neoliberal, business-friendly state recruiting women to manage the problem of garbage, but zero-waste practitioners had a different analysis of the garbage problem, as well as a different set of solutions.

Local Action to Confront Global Problems

Frustrated by failed cleanup campaigns and increasingly suspicious of the nexus between garbage contractors and municipal functionaries, some women retreated from state-sanctioned schemes to form neighborhood collectives for household and community action. Looking to take matters into their own hands, some of them started replacing single-use plastics with reusable materials. Others started learning how to do home composting. They turned to the Internet for advice and learned about a global zero-waste movement (ZWM) that emphasized the principles of responsible consumption: reduce, reuse, resource recovery. They educated each other about the broader ecological impacts of waste dumping, learning that simply cleaning up blackspots was not addressing the root causes of the garbage problem. Some of them visited landfills outside the city, seeing the destruction firsthand. Thus, although many members of Bengaluru's zero-waste management community became involved in neighborhood-level activism because they were looking to clean up blackspots in their neighborhoods, their need to take matters into their own hands also produced, over time, a more ecological understanding of waste flows.

Bengaluru's zero-waste practitioners became concerned about how their everyday consumption practices affected people, things, and places beyond themselves, whether Mavallipura's landfill-adjacent communities or global ecosystems threated by climate change. Inspired in equal parts by globalized "performative environmentalism", which valorizes consumer action for sustainability, and by local state and civil society efforts to restore Bengaluru's garden-city heritage, these movements emphasize the role of individual behavior change and citizen responsibility in addressing the negative environmental impacts of waste.[12] By moving sites of intervention from the streets to within homes, these zero-waste zealots turned to reducing waste, segregating waste at source, and recycling and composting as solutions to Bengaluru's enduring garbage crises. They see place-based collectives such as apartment complex associations or residents' welfare associations as the best sites for action, and enlightened, concerned, and respectable citizens as the best actors to change the status quo.

To achieve zero waste, which includes reducing how much waste one discards, innovatively reusing waste materials, recycling, and composting, community members look to the local, the material, and the manual in place of world-class or high-tech sustainability solutions. Whereas one might expect a middle-class-led effort to emulate other so-called world-class cities in implementing technologies like waste-to-energy plants, communal sustainability practitioners push for homegrown zero-waste management systems that require a lot of manual labor and volunteer effort. Many praise traditional values such as thrift and good housewifery, and promote the use of "indigenous" materials like banana leaves to replace polluting plastics (see also Pathak and Nichter 2021). Indeed, some communal sustainability practitioners could be described as "eco-traditionalist"—anthropologist Gauri Pathak's label for a form of environmentalism practiced by dominant-caste Hindus who frame the Indian (reduced here to Hindu) cultural ethos as inherently eco-friendly, damaged by Westernization, and recoverable through consumer action (Pathak 2021; see also Erler and Dittrich 2020). This eco-traditionalist narrative problematically constructs India's past as a Hindu past to the exclusion of other religious and caste communities—perhaps also explaining why Bengaluru's zero-waste efforts were concentrated in dominant-caste, Hindu-majority neighborhoods.

During fourteen months in 2012 and 2013, I interviewed thirty-eight women and twelve men who had initiated waste segregation programs in their residential neighborhoods and apartment complexes. These programs were concentrated in central, wealthy, dominant-caste neighborhoods like Malleswaram and Jayanagar, or in upscale gated communities in newer, peri-urban areas close to IT parks like Whitefield, Yeshwanthpur, and Sarjapur. Led largely by women who worked in the home, most of these initiatives operated with limited assistance from the municipality. Instead, "passionate" volunteers in the community worked with residential welfare associations as well as NGOs and private vendors to change household waste handling practices.

In the sections that follow, I present some ethnographic vignettes from my fieldwork to show how the leaders of these initiatives utilized their communicative and relational capacities to cajole and convince their neighbors to change waste handling practices.[13] I show how women conducted awareness drives, organized social gatherings, made signage, and even indulged in some old-fashioned face-to-face goading to change the social meanings around waste. They worked with real estate developers and construction managers to purchase organic waste composters and construct sheds for the sorting and storing of dry waste. Finally, they also played a hands-on role in monitoring the day-to-day operations of the program, including carefully scrutinizing the daily discards of their neighbors. Through this, they slowly modified the material arrangements, social meanings, attitudes, and emotions associated with waste handling practices.[14]

The Affective Labors of Sustainability

"Our Life Is in Danger!"

On a Saturday evening in October 2012, I am in KPS Sukh Nivas, a large luxury gated community in Yeshwanthpur, with the Care Collective. Comprising 350 apartments, the complex boasts a tennis court, clubhouse, pool, and even a supermarket and Hindu temple within its walls. We are in an air-conditioned party hall addressing fifty or sixty residents, at the invitation of one of the apartment association leaders. Women again outnumber the men. The invitation was precipitated by BBMP bylaws issued earlier in the month requiring large complexes to segregate waste at source. The bylaws were an outcome of the PIL filed in the Karnataka High Court in July 2012 (see introduction). In response to the demands made by the petitioners, the High Court directed the municipality to mandate source segregation of waste. The BBMP, to demonstrate compliance, started sending out circulars to large apartment complexes and gated communities (termed bulk generators) asking them to ensure segregation of waste at source and in-house composting. It threatened fines for noncompliance. This gated community was composed of new homeowners reluctantly trying to comply with these regulations.

By this time, I had become an honorary member of the group, which took me along for their weekly presentations. We came prepared with props to engage our reluctant and fidgety audience; for example, there are five aprons, each covered with props and images representing different types of dry waste. Today, I am wearing the e-waste apron, replete with CDs, floppy disks, and an image of a tube-light (see figure 2.1). Sheetal kicks off the program. As a "housewife," she rose to fame when a news program featured her organic gardening and composting skills. Unlike Pramila, who tries to strike an inspirational tone, Sheetal starts bluntly: "You have no choice but to comply with the BBMP as there are plans afoot to issue substantial fines to bulk generators who do not manage their waste in-house." Sheetal points to the snacks station set up for the attendees, replete with disposable paper plates and plastic cups, telling them to "stop using these disposable plastics. Buy a set of steel plates and tumblers to use for your parties and events and store it here." The audience fidgets and I hear some murmuring. An older gray-haired man objects: "Madam, the dishes will not be cleaned properly. It might spread disease. We have to be careful."

Chitra, the youngest but most forceful member of the group, interjects, saying, "Uncle, you are old, you remember twenty years ago how we used to wash everything." The older man gets some support from his fellow septuagenarian friend: "But back then we had fewer people! How many

Figure 2.1
Waste segregation presentation at an apartment complex. I'm second from the right.

people are even willing to switch from plastic? People who consider themselves modern will not switch to old-fashioned things, ma." What is left unsaid but can be surmised here is also the influence of caste. Brahminical caste purity-pollution rituals bar the sharing of utensils with non-Brahmin individuals because of fear of ritual contamination. These gated complexes, while class-homogeneous, consisted of people of varied caste, linguistic, and regional backgrounds. Therefore, the idea of sharing utensils is likely to be distasteful to people committed to maintaining caste boundaries. Seeing the conversation sidetracked, Sheetal turns her attention to the women, saying, "Well, I will at least ask the young mothers here. You all will use a lot of plastic while throwing birthday parties for your kids. Instead of doing that, please use reusable." Almost all the women nod enthusiastically as Sheetal smiles at them encouragingly.

In a little while, I walk up to the makeshift stage and perform my spiel about e-waste. After me, another member wearing an apron covered with Tetra Paks and crisp packets comes up: "This multilayer packaging cannot easily be recycled, and we should minimize how much we use of it." The last speaker wears the apron featuring recyclable discards—newspapers, milk packets, aluminum cans, and glass bottles: "These materials are valuable. By collecting them and keeping them separate, you are not only reducing what ends up in the landfill, but your complex will also be able to earn some money which you can use to pay the workers who do the waste collection and segregation." As we speak, I notice the continued skepticism of the audience. When Sheetal comes back on stage to talk about composting, all hell breaks loose. The residents start name-calling, accusing the heads of the apartment association of being impractical. "People won't separate the garbage. The maid will take it and throw it," says one man. "You want us to keep all the dirty food waste in the house?" another shouts.

The head of the homeowners' association, who invited us, steps up with a PowerPoint of his own. He raises his voice as he pulls up a photo of the basement: "Your house may be clean, but look at our basement, look at the kitchen waste. There are rats here! This is a plague. This is happening in our apartment. Right under our noses! Our children go here. Our life is in danger!" During the entire presentation, not even one of the forty women in the audience could get a word in. After the presentation at KPS Sukh Nivas, we were all a bit dejected. Roopa, a woman in her fifties, tells me, "It's usually not like this. Usually, we talk to communities who are already a bit interested. Now, with the BBMP rules, we will have to start going to places like this where people don't care and make them care or make them just do it. It's usually easier to convince the women. They will end up doing most of the work anyway! Those old men . . ."

"That Is a Polite Way of Putting Peer Pressure"

At the Recycling Habba, I meet three women representing their gated community's zero-waste program. Located in Uttarahalli, in the southwestern periphery of the city, Resplendence Godavari was built in 2006. This upscale development has a swimming pool, gymnasium, outdoor play area, rainwater harvesting system, 24/7 security, and backup power. A two-bedroom apartment in this community cost 7 million rupees in 2020.

Manisha: Can you tell me a little bit about your program?

Geetha: Few months ago, we got a notice from the BBMP saying that segregation is soon going to become mandatory. So that was what we used to wake up the community saying we better form the habit before it gets mandatory, so we do not face problems later. So, we began looking into the model that is going to suit our community . . .

Manisha: So how did you start that process? Who did you contact to put this model together? Did you know of other programs going on?

Geetha: At that point we researched on the Internet and went on the TCGF website. And we found a person there who was a cousin of one of our community members. And we went and talked to her. She told us dry waste segregation is the best way to go first. And we did more research and talked to people in the city. But then we ourselves decided the best model for our community. We have a big apartment of 136 flats, just two to three people cannot do it. So, we have a team of twenty-four.

Manisha: Was it easy to build that team?

Geetha: First, the core team was formed: four passionate people who wanted to make a difference. We visited some complexes where waste is being segregated. And then actually we made our own process because every community is different. Then the psychology comes. We selected people who have a presence in the community, who have a social group by some way. If you have about ten friends, you can make a peer group and then you can create pressure. So, select such ladies who make the difference at the floor level. Make them as the floor representative, so they can convey the message.

At that point, a man walking by interrupted the interview, butting in to ask, "So what are you all doing, segregation, recycling?" Geetha replied, "Yes, segregating—recycling is done by a third party." The man clarifies, "So all you are doing is segregation?" She laughs and responds, "Yes, that is all we are doing." The second lady, Sanvi, takes offense: "It's not all that we are doing, it is a lot that we are doing! Recycling, there many companies who are doing, but segregation is the necessary step. You have to change in the house first."

Geetha (cont.): We built up a momentum. We made a PPT. We started off talking about why it is necessary. We gave the presentation to the block reps. Then the block reps gave it to the floor reps. Then the floor reps did it on the door-to-door. Floor reps trained the community members. So, there are three iterations happening on the same information. The community, by the time we kicked it off, knew all about it.

Sanvi: Yes, I think we even had three sessions, three parties, where the floor people would come together and get educated.

Manisha: Once you started, if a house is not doing the segregation, what do you do? Does that happen?

Sanvi: So actually, there are very few apartments which are refusing. We have gone to them and as much as possible, we have explained to them. There are some people who have said, milk packets we cannot wash and dry and give every day. So, we said, okay, at least the rest of it, please do. And if there have been things that people have been putting in dry waste that they should not, like sanitary napkins, then we give them feedback.

Manisha: How do you know who is not segregating?

Sanvi: For about two months, the floor reps and block reps recorded the segregation every day using this sheet. [She shows me a sheet featuring a segregation log with each row pertaining to an apartment number. The columns indicate whether dry waste has been kept out and whether the segregation has been done according to the specifications provided by the organizers (see figure 2.2).] So we have been going to people and telling, we know you are mixing the waste. Are you confused? How can we help? And there was a peer pressure created because they knew we were recording. They knew which house was doing it and which one was not. So, people fall in line.

Geetha: That is a polite way of putting peer pressure. No policing. Telling them that we are also along with you. Floor reps were given a handout, as to

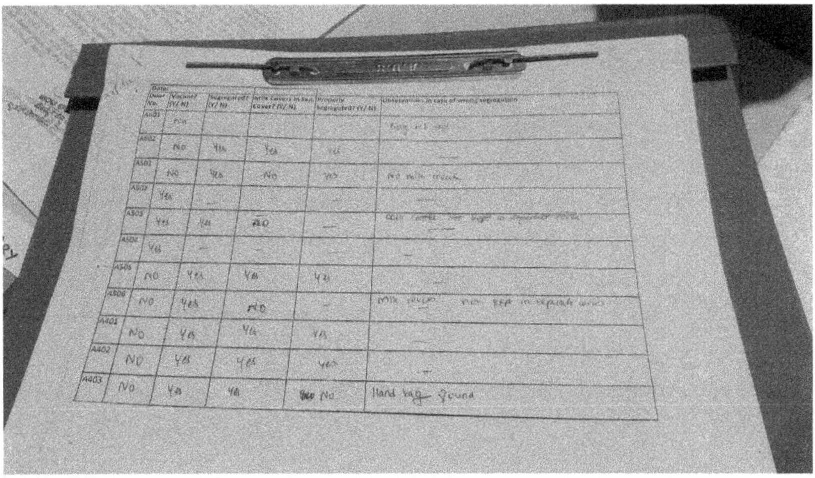

Figure 2.2
Log sheet used by the floor leaders of the zero-waste community to monitor segregation of household waste. The notations indicate improper segregation.

when they go to the door-to-door on how they should talk. It should not look like preaching. We talked to people. We researched. We educated ourselves.

Concerted affective labor, a form of socially reproductive work, is engaged by women leaders of zero-waste initiatives to rearticulate the meanings, associations, and practices of waste and its management. Many middle-class households have traditionally segregated high-value recyclable waste or *kabad* to sell to *kabadiwallahs* (local scrap shops) as a practice of good housewifery, but these programs demand more from households by asking them to segregate all the waste produced in the home into dry, wet, and sanitary waste. Thus, it was no longer enough to keep the paper or glass aside. Rather, the cook now had to make sure that vegetable peels did not end up in the dustbin with the cellophane wrapper. The maid who swept the floor and cleaned up after the family had to remember not to put the hair from the drain in the same bag as a chocolate wrapper discarded by a wayward child. The woman of the house had to supervise all of this, carefully scrutinizing her paid help while also changing her own practices and convincing her husband that it was safe to warehouse discards for up to a week.

Affective labor is deployed to manage the emotions of others in relation to a particular task.[15] Emotions play a key role in reinforcing social norms around cleanliness, hygiene, and proper conduct within the home (Saha-kian and Bertho 2018). Inducing feelings of shame, anxiety, or blame has been used to maintain social control over how practices are reproduced. In a social milieu marked by caste, emotions of disgust, distaste, or disinterest characterize how dominant-caste Indians relate to waste. On the flip side, those working with waste are marked by toxic exposure and caste-based humiliation (Kornberg 2019a). Reordering these associations requires effort that goes beyond a single presentation or flyer.

Zero-waste leaders direct the affective labor of energizing, cajoling, con-vincing, gently censuring, and even shaming their neighbors to reorder emotional and practical relationships to household discards. Such affective labor undergirded every zero-waste initiative I encountered; it served to break habits and created new ones. Leaders helped validate and legitimize activities that were not previously seen as modern or sanitary with new say-ings, which generated new regimes of meaning as well as new ways of doing things. Through these new meanings of ecological and civic responsibility, including sayings like "It starts at home," household segregation practices circulated in neighborhood-based collectives and citywide networks, form-ing communities of belonging and social learning.[16]

My ethnographic examination shows that community serves as a site of social learning because of the care work of its members. Self-made zero-waste leaders exercised care in translating scientific concepts into meta-phors of home life for their audiences. For example, during one awareness session I attended, Pramila said, "We segregate everything in our homes—clothes, our spices in the *anjana potti*—so why not waste?" Here, Pramila mentioned the *anjana potti*, a staple in most middle-class kitchens that is used to store spices, to suggest that women like her were already used to sepa-rating things in an orderly fashion—alluding again to an eco-traditionalist ethos. Similarly, Sheetal, who is able to articulate scientific explanations around bio-geochemical nitrogen cycles and distinguish aerobic and anaer-obic composting down to its chemical minutiae while talking to me (an erstwhile botanist), emphasized the sensory details of smell and feel and the intricacies of practice when addressing an audience of fellow housewives: "When you start composting, there will be some maggots, but don't worry, and don't tell your husband, he will freak out. To avoid the maggots, all you have to do is add some sawdust. If it is smelling, just add some sawdust.

Anyway, maggots are not bad, they are also natural. And if you have any issues, you can always call me."

The narratives used by these women reflected a shared experience of middle-class womanhood—preoccupied with cooking, managing husbands and children, and provisioning the home. Once, a practical question from a woman in the audience about whether storing dry waste for a week would attract cockroaches prompted Sheetal to offer some "tips" on how she cleans plastics that have been contaminated with food items. She had generated these through trial and error: "With your milk packets, just cut it across the seam and then rinse it with some water and make tea with that little milk. For a plastic packet with oil, cut it and roll your ball of *aata* [flour] in it that you can make *chapathis* [flatbreads] with later in the day. That way all the oil and milk is gone and you can safely put it in the dry waste." Sheetal, who is self-taught, is communicating the practices of zero waste using examples that are familiar to her audience. To do this, she draws on her identity and experience as someone who actually does the work of running a home on a daily basis as a homemaker. She validates the fears of her audience that their home will be overrun with vermin if they store their waste and finds ways to reassure them. She talks about trivialities in banal detail, but in doing so, speaks from experience and empathy.

Attempts to create new material arrangements, norms, and meanings around waste handling were not always successful, and I encountered a few zero-waste management initiatives that had been disbanded because of resistance from some residents, often related to concerns about health and aesthetics. In other cases, zero-waste management initiatives that had been set up with great enthusiasm fell into disrepair in the absence of monitoring by resident volunteers or because the core initiators had moved away. Speaking with these women, it became clear to me that the success of these programs was highly dependent on the ability of leaders to convince and compel their neighbors, monitor compliance, and keep everyone energized and engaged. It was a lot of work.

Exhausted at the end of a long day visiting far-flung gated communities, I often wondered how these women stayed motivated and interested. As I spent more time with the Care Collective and attended waste management festivals, it became apparent that volunteers derive strength from each other, generating good feelings about being in like-minded community. At another recycling festival in March 2013, Devaki, who was a leader in a gated complex in Indira Nagar, decided to join the Care Collective team for lunch. As everyone chatted and shared stories, she remarked, "Ah, I am so happy to be here with you all, you understand me, and why this is important." Indeed, when faced with pushback and criticism from family members, neighbors, and most notably husbands about their "dirty" pursuits, people like Devaki turned to events and online networks on Facebook and email groups such as Zero Waste Bengaluru for support and encouragement.

These events and social gatherings began to serve as spaces for interaction rituals. The sociologist Randall Collins (2014) has argued that interaction rituals generate emotional energy or "collective effervescence" under certain circumstances, which in turn leads to a repeat of these rituals. At these zero-waste festivals, volunteer leaders from across the city gathered with their advocacy materials, decorative bins, and other props in hand. In meeting with one another, they validated one another's commitment to the cause: they knew how to clean up the city, protect the environment, and even be a little thrifty in the process. They shared tips and stories of success and frustration over apathetic families or stubborn neighbors. When zero-waste ladies gathered, they lifted each other up, steeling themselves to continue the work back home. With their emotional energy replenished, they took on the task of cajoling their neighbors and eventually lobbying the state.

Cultures of Servitude in Communal Endeavors

Swamy describes the zero-waste project he volunteers with in his upscale neighborhood:

> We started piloting on one street that had forty homes. We give the houses two bags, one for plastic and paper, another for e-waste and glass. Once a week a guy, who we have hired for the job, collects it. But we didn't want the segregation program to be something that was just run by us, we wanted it to be something that we did physically. Because there's a stigma attached to handling garbage, and nobody wants anything to do with it. Caste thing. And we didn't want it to be— oh I want to get rid of my garbage and that's why I participate in it. We wanted people to get joy out of it, the joy of feeling that they are personally doing something for the environment. So, we used to actually walk once a week on collection days with the collection man, door-to-door. Ring it, say hi. And then take their

bags, put our hands into the bags, just to check if things are good. But mostly it was maids who did the sorting.

As the only man in the core group of ten leaders, Swamy talks about challenging the "caste thing" by accompanying the man they hired to collect segregated waste during his collection rounds. What Swamy is referring to here is the purity-pollution dynamic of caste. Dominant-caste Indians, and in particular Brahmins like Swamy and myself, are raised in a habitus that sees the handling of many types of waste as a task that marks the body as impure. Brahminical life is marked by a proliferation of ritualistic cleaning tasks combined with an active distancing, distaste, and stigma for pollution, as well as the social groups who perform these so-called polluting tasks. This capacity to distance derives directly from the social ecology of caste, through which dominant castes force Dalits and other oppressed castes to perform polluting occupations, so that the former do not have to expose themselves to bodily and other wastes (P. Gupta 2022; Guru 2011). Thus, Swamy saw putting his hands in someone else's waste as a caste-transgressive act.

Generally, however, the waste management initiatives I profiled relied on workers to collect, clean, and sort waste. Many neighborhood systems employed workers to segregate and transport waste, just like middle-class urban gardeners who had off-site gardens employed full-time staff to tend their plots (van Holstein 2019). In the swanky gated complexes I visited, the housekeeping staff who maintained the common areas were also now tasked with waste sorting. In residential subdivisions, contract sanitation workers took on this additional burden. In some cases, organizers supplemented the salaries of waste workers with money recouped from the sale of recyclable material, while in other situations, there was no commensurate increase in pay. A few gated complexes outsourced the job, hiring waste management service providers to carry out secondary and tertiary segregation.

Communal sustainability operates within a rigid caste order that constrains the role and agency of waste workers, casting them as noncommunity members who need to be disciplined and monitored. For instance, during my visits, the housekeeping staff, the sanitation workers, and the maids never spoke; instead, they were spoken to. Middle-class and/or dominant-caste project coordinators who speak English and Hindi expected waste workers, who were usually lower-caste and poor people, to follow orders and obey rules (see figure 2.3). Workers spoke in Tamil, Telugu, or Kannada,

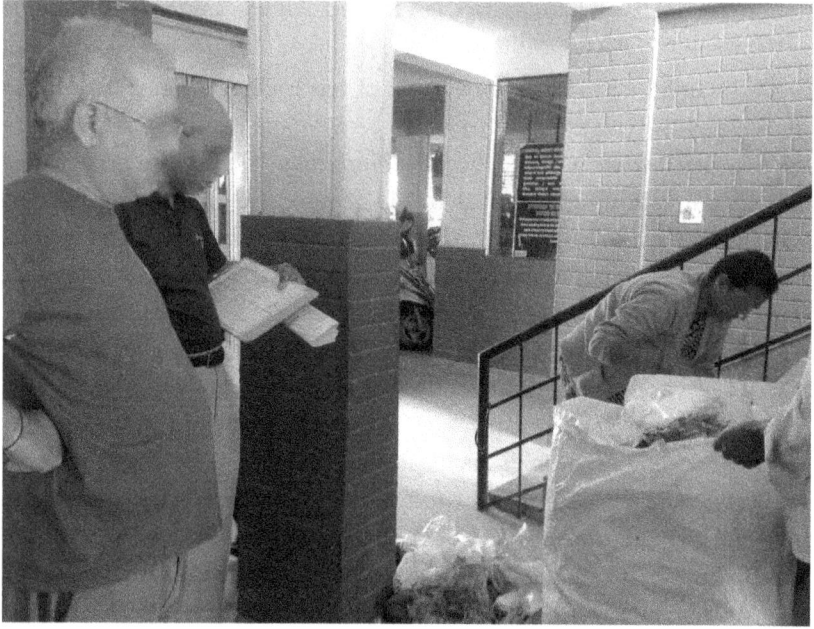

Figure 2.3
Retired middle-class men supervising housekeeping staff. The two men are coordinators of a zero-waste program; they are watching over a female worker as she sorts
through a pile of imperfectly segregated discards. This picture captures cultures of
servitude in action.

the languages of waste work in Bengaluru. The following vignette from my
research notes offers a glimpse of this dynamic:

> I arrive at the gated complex in East Bengaluru. It is located in one of Bangalore's
> most expensive and posh localities. The complex is a large development consist-
> ing of over a thousand upscale residential apartments and some corporate offices.
> I am scheduled to meet with Mira, who spearheaded the community's zero-waste
> project called Project Green X. I meet her in the parking lot of one of the large
> apartment blocks (of which there are sixteen), and she quickly ushers me toward
> a cart being pushed by two young women in green uniforms. "You're a little late,
> but just in time to see the collection," she says. We follow the cart being pushed
> by these two quiet women to the shed where we see two other women sitting
> inside, amidst ceiling-high piles of papers and plastics, sorting through the refuse.
> Mira explains to me that the women are employed to do secondary and tertiary
> levels of segregation, where paper is separated from plastics, and high-value items
> like milk packets, shampoo bottles, and glass are set aside. We are joined by a young

man who is introduced to me as the supervisor of Project Green X. He speaks harshly to the women, telling them to speed up their work. It is his job to make sure that the whole operation runs smoothly. My guide tells me that the salaries of the supervisor and the women segregating waste are paid with the money the complex earns from selling the recyclables to different vendors.

Leaders of neighborhood initiatives sometimes accused waste workers of having self-interested motivations or other vested interests. They were not to be trusted to do the right thing for the planet. As one woman, Prema, told me, "Once they (the workers) finish sorting the recyclables, we put them in this room and only I have the key. We fund this program by selling the dry waste. And if we leave it out here, they will steal all the good stuff." If the workers were indeed to "steal" the recyclables, they would be selling it to a scrap dealer for money, who in turn would divert it to a recycler: all key lynchpins in an informal recycling economy that saves municipalities' money and mitigates carbon emissions (as we will see in chapter 3). However, for middle-class actors, it was important that they were the ones diverting waste from the landfill and that they retained full control over the process.

By refusing to recognize oppressed-caste and lower-class workers as ethical actors in the same light as middle-class housewives, and by denying their own self-interested motivations in wanting "clean and green" cities, communal sustainability gives material form to performative environmentalism's discursive and aesthetic biases. Further, eco-traditionalist ideas that imagine a Hindu India as inherently eco-friendly deny the role of caste-based labor in maintaining environments (M. Sharma 2017b). These ideas also further marginalize non-Hindu communities in environmental politics. Thus, if a middle-class person separated trash in their home, deposited their recyclable waste with a scrap dealer, or composted green waste, they claimed these as acts of ecological stewardship. If waste workers within the home or in the neighborhood carried out the same action, they were just doing their jobs or stealing.

Reproductive Labor as a Route to Environmental Politics

In the previous sections, I have argued that gendered and casted reproductive labor within the home, combined with affective labor in the neighborhood, create the social infrastructures of a "clean and green" city. However, I caution against a reading of these initiatives solely as the successful expression of governmentalizing projects led by the state. Rather, participation in community-based zero-waste projects becomes a route through which many women and workers engage in urban politics in the city—liaising with local councilors, municipal functionaries, and eventually senior bureaucrats and the courts. This is unpaid care work, but also a mode of civic engagement and ecological citizenship. Communal sustainability is a paradox—it can exacerbate existing class, caste, and gender-based oppression while simultaneously opening new avenues for political participation by middle-class women and, as we will see next, by some waste workers.

Quality of life and urban sanitation issues have long been a route through which women, in their roles as mothers and carers, have engaged in urban politics (Macgregor 2007). Sherilyn Macgregor argues that these activities are not solely expressions of care work, but also instances of women, a group marginalized in both urban and environmental politics, staking claims to full participation in remaking the public. I find too that, despite adopting an apolitical framing of community, these zero-waste collectives, led by and largely composed of women, end up doing politics.[17] Their dominant-caste positions, practical orientations, and homely metaphors enable them to position themselves as neutral mediators of what is best for the city and the planet.

While Bengaluru's zero-waste leaders might discursively eschew politics in favor of commonsense local action, they still engage in activism. Emboldened by local successes, leaders of these zero-waste management initiatives formed citywide coordination groups like the Clean and Green Forum (TCGF). Groups like TCGF have facilitated the replication of successful systems in new localities by organizing awareness campaigns and connecting households with waste service providers from the private and nonprofit sectors. TCGF claims that more than fifty thousand households are participating in grassroots waste segregation programs. Volunteers who effectively straddle the scientific with practical and experiential knowledge and who maintain community efforts through affective labor are critical agents in this transformation.

In 2012, believing that the time was right to intervene in municipal infrastructures, TCGF resorted to one of the most tried-and-tested tools in the middle-class civic participation playbook—filing a public interest litigation (Bhan 2016). Citing local successes as evidence of the efficacy of their proposals, they forced changes to municipal waste management codes to mandate source segregation, recycling, and composting (Lutringer and Randeria 2017). On the ground, they successfully infiltrated the municipal machinery, building convivial relationships with key officers in the BBMP Solid Waste Management cell, as well as with ward-level sanitation engineers. The policy changes achieved in Bengaluru, such as mandates for source segregation of waste and requirements for bulk generators (like hotels, apartment complexes, and hospitals) to manage organic waste in-house, have now been enshrined in the 2018 National Municipal Solid Waste Management Rules. Importantly, and unlike neoliberal programs like the Smart Cities missions, which pervert participatory processes and cherry-pick citizens' voices that are well-aligned with the official technocratic vision of "global smart urbanism" (B. Ghosh and Arora 2022), Bengaluru's women-led ZWM has helped slow the powerful forces pushing privatization and incineration for waste management.

The leaders of these sustainability projects are surprised at their own success. In 2013, I was driving back with Savitha, a prominent zero-waste advocate, from a court hearing in which the judges had just declared that Bengaluru must set up materials recovery centers in each ward. When I asked Savitha how she felt about this major victory, she said to me, "I am so surprised. Who are we to tell the government what to do? A rag-tag group of housewives!" In reality, this group frequently tells the government what to do and sometimes the government listens. Three years later, in a follow-up interview in 2016, Savitha elaborated: "Our power comes from our ability to talk about the practice and not just the theory." These successes demonstrate that care work convened through affective relationships can be a potent force in propelling behavioral and even policy changes.

This case study of Bengaluru's zero-waste communities shows stratification and diversity in Bengaluru's cadres of urban reformers. Although zero-waste networks are largely middle class, they are qualitatively distinct from the male-dominated, corporate-elite-led reforms of BATF. It also points to a key instance in which there are ruptures between state-led and corporate-backed governmental agendas of cleaning up the city through privatization, spectacular infrastructuralization, and mechanization, and a bottom-up movement emphasizing community involvement and vernacular solutions. These are expressions of what Arjun Appadurai calls governmentality from below (2001), which in some cases dovetails with top-down governmentalizing agendas and sometimes challenges them.[18]

These new socio-material infrastructures, with their metabolic dependencies, also open a window to transforming sustainability politics in the city from a politics controlled by dominant-caste male technocrats to one that has more diverse representation and participation. Bengaluru's communal sustainability proponents emphasize personal responsibility, accountability, and civic duty. "Bourgeoisie environmentalists" (Baviskar 2011) deny their metabolic dependence on the working poor for the reproduction of their lives in the city. In contrast, zero-waste proponents acknowledge and rely on domestic maids, housekeeping staff, sanitation workers, and other waste workers to enact the tasks of zero waste. In doing so, the ZWM leaves in place the structures of caste and class that normalize subjecting some social groups to toxicity, risk, and humiliation to subsidize comfortable and safe lives for dominant-caste and socioeconomically privileged groups. Yet, it also opens discursive space for new kinds of political negotiations, now and in the future.[19]

Communal sustainability's emphasis on the local, material, and manual creates a dependency on the labor of waste workers within the home, in the neighborhood, and in municipal systems. Do these metabolic dependencies on manual environmental labor create new avenues for some sections of the waste-engaged poor to participate in urban environmental politics? In the chapters that follow, I tell the story of how an organization representing informal waste pickers—a structurally oppressed group whose already marginal and precarious livelihoods were threatened by the zero-waste agenda—leveraged communal sustainability to defend waste-pickers' right to waste and make claims for inclusion.

3 Entrepreneurial Environmentalism

In an already sweltering February heat in 2013, the who's who of Bengaluru's waste management world gathered for an expo called Wake Up Clean Up (WUCU) in Freedom Park.[1] The event was a collaboration between the Bruhat Bengaluru Mahanagare Pallike (BBMP) and an elite civil society organization called City Connect. City Connect bills itself as a platform uniting business, civil society, and government to address "livability" issues. It is your typical corporate-friendly civic organization, composed of executives from Bengaluru's IT and biotech industries.[2] According to City Connect's "Chief Evangelizer," V. Ravichander, WUCU could help convince big real estate developers and corporate honchos that managing waste at source was "not just a responsible, do-good action but a wise business opportunity as well."[3] City Connect predictably highlighted private-sector involvement, business acumen, and techno-rationalist approaches, in keeping with neoliberal sustainability's business-friendly ethos. Each morning, the event started with a power *nashta*, an exclusive catered breakfast for Bengaluru's top corporate leaders, who emerged like ants from Bengaluru's special economic zones.[4]

Designed and coordinated by elites, with panels and programming taking place almost exclusively in English, the event sidelined input from labor groups and even middle-class community organizations.[5] Organizers finally asked *guttige pourakarmikas* (contract sanitation workers who do door-to-door collection of waste) to appear on day four. They were bussed in by their employers. Government officials told them that, with the new rules, they should collect only segregated waste from households and never mix dry and wet waste. When I checked in with two *pourakarmikas* after the session, they said in Tamil, our shared language, "These people live in a dream world. I cannot refuse to pick up the mixed waste. They [the residents] will shout at me and call me names, and the supervisor will be angry. And even if one household mixes it, and I only have two bins, one for dry and one for wet, where am I to put the mixed waste? Then I will have to mix it too, right? They don't understand anything."[6]

Both start-ups and established companies peddled their wares at the event, selling everything from organic waste composters to labor services for waste management. One new start-up offering recycling services had even named itself Ragpicker, perhaps as a tribute to the city's waste pickers, itinerant buyers, and scrap dealers, who, until recently, were the only actors reclaiming resources in Bengaluru. There were a few old-school waste pickers at the event as well, representing Parisara Tanda, a newly formed waste-picker organization in the city. Parisara Tanda insisted on having a presence at the event because new recycling schemes could dispossess the city's fifteen thousand waste pickers if reclamation and resale rights for recyclables were granted exclusively to private companies under the pretext of diverting these materials from landfills.[7] This posed an existential threat to the informal waste economy by restricting access to waste and further criminalizing this work.

Wearing new uniforms and gloves, Periamma and her sister, waste pickers affiliated with Parisara Tanda, weaved purposefully through the crowds. They worked in the background, shepherding wayward discards into ubiquitous blue bins, hiding the waste generated by the thousands of people attending the expo. Recycling signage appeared to perplex visitors who were unaccustomed to segregating their waste. Periamma had to resort to rummaging through bins, fishing out paper food containers, plastic bottles, and soda cans from among banana peels and wasted food. She joked to me, "Isn't it funny that they are not keeping the waste separate in a big, fancy event dedicated to waste separation! This is bigger than a wedding!"

In the context of an elite-dominated environmental agenda that denies ecological legitimacy to the poor and favors privatization and marketization as solutions to environmental problems, how do groups who are threatened by elite-driven sustainability initiatives protect their livelihoods? And what does an examination of the claims-making strategies utilized by excluded constituencies tell us about the ways in which dominant urban sustainability discourses constrain or encourage possibilities for involvement and action by highly marginalized groups? In this chapter, I draw on my decade-long engagement with Parisara Tanda, an organization that represents waste pickers in Bengaluru, to demonstrate how, in an era of "eviscerating urbanism," claims-making strategies evolve beyond ethics or even moral arguments to instead focus on utility.

Parisara Tanda forefronts the environmental and economic usefulness of waste picking, fashioning waste pickers as entrepreneurial environmentalists. They pragmatically deploy a claims-making strategy that, instead of emphasizing ethics or human rights, highlights waste-picker utility in providing environmental and economic services to the city. By essentializing waste picking as an act of silent environmentalism, they demand ecological legitimacy for waste pickers. They simultaneously highlight the robust entrepreneurial abilities of waste pickers because, in cities like Bengaluru, entrepreneurship is celebrated. The environmental entrepreneur thus emerges as the ideal neoliberal subject who not only takes care of their own needs, but also produces surplus environmental and economic value for the city and the economy.

Making Claims in Eviscerating Urbanism

In the twenty-first-century neoliberal city, where market supremacy and private enterprise rule, specific subjectivities are venerated and valorized. Entrepreneurs and consumers are celebrated because the social relations implied by those categories are transactional and convened in the market. Workers and citizens, on the other hand, demand social relations beyond the market that in liberal states include obligations and entitlements such as rights from the state or welfare benefits from employers. Workers and citizens, particularly propertied ones, demand acknowledgment and management. Finally, the poor are the most problematic urban subjects, as they demand what appears most scarce in the neoliberal city—an ethical engagement.

The neoliberal state in India is increasingly reneging on its postcolonial development commitments to fight poverty and improve the living conditions of the dispossessed. Instead, as geographers Vinay Gidwani and Rajyashree Reddy (2011) explain, in late capitalism, ethical engagements with the poor are substituted with revanchist politics, at worst, or ad hoc relationships characterized by abjection and patronage, at best. This "eviscerating urbanism" relies on and reproduces continuous processes of classification, categorization, and stigmatization between the deserving and undeserving poor; between legitimate, rights-bearing citizens and the illegitimate, burdensome precariat; between legal and illegal land uses. Categorization justifies attending to the claims of some "worthy" social groups, while marking racialized others for suffering, humiliation, and even premature death. These processes of categorization are relational, historically anchored, and articulated with the biases of coloniality, race, caste, and class.[8]

I became aware of these categorizations early in life as a child raised in a Brahmin, middle-class family in Chennai. I internalized them as normal. My family's class and caste position afforded us access to a domestic helper, a "servant," who cleaned our home and transported our refuse to the dump at the street corner every day. Yellamma was never allowed to drink water from cups used by my family. She was refused entry into the prayer room. It was only years later that I recognized these daily humiliations as intimately connected to the workings of caste and as manifestations of my parent's desire to maintain Brahminical lines of purity and pollution.[9] Thanks to Yellamma, I never had to deal with my waste.

Similarly, my encounters with and impressions of waste pickers stretch back to my childhood. My father, like many parents, devised innovative ways to get five-year-old me to eat my meals. While children in the United States are told about poor starving African children to motivate them to not waste when others are left wanting, my father's approach was a little more coercive. He used to tell me that if I did not finish what was on my plate, the man we occasionally saw at the streetside dump near our house would carry me away. This man, who carried a brown hemp sack and wore clothing that looked tattered to my young eyes, spent early mornings rifling through the bin at the street corner. My father's casual threat that I would be spirited away in that sack, along with all the waste, played into my already emerging fears of the lone male wandering the streets. My worldviews were formed by my Brahmin family who, through their words and actions, normalized our distancing from those who worked with waste, as well as people who ate meat or worked in occupations like leather tanning. At the same time, I received no critical education about caste in my schooling. My books simply detailed the *varna* system and asserted that the caste system had been abolished when India became independent. The ongoing realities of caste oppression were obscured. These ideas influenced my own engagement with waste pickers, until I started to recognize my prejudice and began the lifelong process of unlearning how I had normalized, sanctioned, and participated in caste-based oppression.

Waste pickers make a living by reclaiming value from waste. By gleaning recyclable and reusable materials such as plastics, glass, paper, and metals from discards, they reduce costs for municipalities and regulate urban environments (Chandran et al. 2019; Gutberlet et al. 2017). Waste pickers and informal recyclers together process 20–50 percent of the waste generated in the cities of developing countries (Wilson et al. 2009). The middle-class habitus that families like mine rely on to reproduce caste privilege would fall apart without the infrastructural, metabolic labor carried out by waste pickers and other waste workers who touch and remove the refuse expeditiously expelled from homes like ours. As Vinay Gidwani poignantly puts it, waste pickers are "thrust to the margins of the contemporary city and face daily humiliation within it, but without [their] anonymous labor and artfulness, the urban fabric that city dwellers take for granted would rapidly unravel" (2013b, 774).[10]

The toxicities of waste, combined with the stigma and abjection it materializes, expose waste pickers to symbolic and physical violence in the city. The persistent stigma and hazard of waste work—work in which individuals who hail from Scheduled Castes and Scheduled Tribes are overrepresented (Lee 2017)—is itself tied to the workings of caste as a long-standing, pervasive regime of social control, oppression, and humiliation in South Asia. As best explained by Dr. Bhimrao Ambedkar, India's preeminent Dalit revolutionary, anti-caste scholar, and chief architect of the Indian constitution, "the caste system is not merely a division of laborers—which is quite different from a division of labor—it is a hierarchy in which the divisions of laborers are graded one above the other" (2014, 234). Further, "caste is a social construct that stratifies human beings and their humanness" (Ranganathan 2021, 5). Brahminism justifies relegating some social groups to hazardous, toxic, and dangerous work because it establishes an ontological bond between waste occupations and Scheduled Caste status (Swaroop and Lee 2021). Therefore, those who work with waste are usually assumed to be Scheduled Caste or Dalit (even if they are not), thus subjecting them to caste-based stigma (Kornberg 2019a). This stigma marks waste pickers as disposable people even while their labor is essential to urban social reproduction, just as it conscripts municipal sanitation workers into "death work" under the ruse of secure employment (Lee 2017; Sreenath 2019).

Casteism, like racism, while legally outlawed, is still ubiquitous. In the twenty-first-century Indian city, it is often conjoined with and cloaked in neoliberal stigma. Communication scholar Rebecca de Souza (2019) defines neoliberal stigma as a narrative that identifies individualism, hard work, and personal responsibility as defining attributes of dignity and citizenship. Neoliberal stigma demotes those who do not live up to these parameters to the status of the unworthy or dangerous, blaming them for their situation of poverty and precarity. It also takes attention away from the structures of injustice and oppression that produce poverty and inequality. Because waste pickers belong to oppressed castes, their labor is taken for granted by oppressor castes and deemed undeserving of adequate compensation. Migrant waste pickers are rarely recognized as legitimate citizens of urban polities, especially if they also belong to religious minorities (Chu and Michael 2019). Waste pickers are thus making claims in a context that denies not only their economic worth, but also, more fundamentally, their rights of citizenship and even their humanity.

To explain the strategic discursive choices that Parisara Tanda made in naming and framing waste pickers in such a context, I turn to Gayatri Spivak's concept of strategic essentialism. Spivak coined this term to recognize and name the challenge facing those who seek to represent the subaltern for political purposes (Pande 2016; Spivak 2010).[11] According to her, a strategic, as opposed to substantive, essentialism recognizes the impossibility of an essential subject position while at the same time acknowledging the necessity of some kind of "strategic use of positivist essentialism in a scrupulously visible political interest" (Spivak 1988, 13). Essentializing involves the smoothing over of difference, or what anthropologist Dorothy Hodgson calls "strategic silences," and the projecting of shared identity (Hodgson 2011). For Spivak, strategic essentialism is a pragmatic tactic to be provisionally employed for political action (1999). Once equal rights have been achieved, the essentialized identity becomes superfluous. Strategic essentialism has been used as a tactic for political action by feminist, LGBTQ, and multicultural movements that construct essentialist categories (such as woman, gay, people of color) and mobilize them to pursue transformations that can deconstruct and undo those very categories that determine access to power and rights in society. It has also been used by some rural and indigenous environmental movements to gain sympathy and support from urban elites in India (Baviskar 2005).

I argue that in advocating for waste pickers, organizational interlocutors, many of whom are middle class and do not belong to oppressed castes, engage in strategic essentialism to "clean up" waste-pickers' reputation in the city. Just as the city is cleaned and greened using ecological and neoliberal language, so is the waste picker. Are the political interests acted on scrupulously visible? To organizers of waste pickers, yes. The threat was imminent: the communal sustainability agenda was threatening to dispossess waste pickers of their livelihoods by turning municipal waste from an urban commons to a commodity frontier controlled by private companies. The claim then became that waste pickers have a right to waste. However, the tactic to achieve the claim became one that downplayed the notion of rights and instead focused on waste-picker utility and usefulness. It became about identifying why waste pickers were worthy of access to waste, worth that was established as emerging from their essence as silent environmentalists and robust entrepreneurs.

Formalization, Privatization, and Waste Resourcification

To comprehend these claims-making strategies, we must first examine the city's waste management landscape and the ways in which formalization and privatization schemes shape waste picking and other forms of waste labor. Across cities in Asia, Africa, and Latin America, waste pickers do their work on the street, at the dump, in the transfer station, and the informal settlement where they live. In each of these spaces, their ability to glean a livelihood depends on brokering access to territory and materials. They negotiate with a diversity of urban actors: municipal functionaries, waste contractors, sanitation workers, police, resident's welfare associations, and sometimes street dogs and pigs. Waste pickers often lack secure housing, identity documents, or access to other social safeguards. Often denigrated as unorganized or ad hoc, their work is in fact purposeful and planned, from decisions around when and where to pick, to classifying materials and deciding how much to sort and store before selling.[12]

Waste picking is racialized and gendered work, often carried out by minoritized social groups. Bengaluru's waste pickers and sorters are mostly Scheduled Caste and Scheduled Tribe, and a majority of them are women between the ages of thirty and fifty.[13] Refugees and migrants who lack official documentation and employment opportunities often turn to waste picking to survive, even in the Global North (Bonatti and Gille 2019). A growing proportion of Bengaluru's waste pickers are Muslim migrants from West Bengal. Waste recovery work also involves carefully policed caste differentiation, especially between dry and wet waste work (Harriss-White 2017). Women from the most oppressed castes go through wet unsegregated waste, pulling out materials like plastics and metals, while itinerant buyers, usually (higher-caste or Muslim) men on bicycles, buy high-value dry waste like paper and glass directly from households.

Waste pickers and itinerant buyers sell recyclable materials to scrap dealers and aggregators. Indian cities are dotted with scrap shops where small-scale traders called *kabadiwallahs* buy, sort, and store scrap such as paper, glass bottles, or aluminum cans. Scrap shops amass larger volumes and then resell these materials to wholesalers and recyclers inside and outside the city (Gill 2009). Recyclers often operate in urban peripheries or informal settlements, where they use a range of manual and mechanical processes to convert plastics and metals into inputs for manufacturing (Doron and Jeffrey 2018, 129). These chains of resource recovery are hierarchical and relational; relationships of solidarity, competition, exploitation, and accommodation coexist. For instance, scrap dealers sometimes employ waste pickers as sorters, give them accommodation, and provide them with loans during times of need. At the same time, Dalit women waste pickers are susceptible to harassment and exploitation by higher-caste male scrap dealers (Sonal 2023, Wittmer 2022).

Dominant urban and environmental policies dismiss these systems of resource recovery as informal, inefficient, and polluting (Corwin 2020; Guibrunet 2021), and the individuals who do this work as antisocial or abject (Millar 2018). Classifying waste picking as informal work further renders this infrastructural labor invisible and even criminalizes it. Broadly, the term informal is used to refer to economic activities that are not directly sanctioned, regulated, or recompensed by the state (Banks, Lombard, and Mitlin 2020). The global narrative on informal work as articulated in the Sustainable Development Goals defines it by what it lacks. This narrative implicitly, and sometimes explicitly, devalues nonwhite, nondominant-caste bodies. Racialized groups like Dalits in India, Afro-Brazilians in Brazil, or Mexican immigrants in the United States are more likely to engage in undervalued forms of work, such as care and informal work.

Informal work is contrasted with formal work, which is defined by the International Labour Organization as waged work protected by labor laws guaranteeing fair compensation. However, this binary ignores the fact that formal, waged work is more the exception than the rule (ILO 2018), and that characteristics associated with informal work—low pay, job insecurity, and temporary, contract-based employment without benefits—are becoming generalized (Breman and van der Linden 2014). On the ground, moreover, the sharp distinction between formal and informal work is hard to spot (Harriss-White 2017). Formal-informal relations in waste range from coordination to interdependence.[14] As we will see, neoliberal disinvestment has further blurred boundaries (Miraftab 2004b).

A brief look at the history and evolution of municipal waste management systems in India illustrates this point. These systems, set up under the British colonial administration, always depended on manual labor and informal arrangements, locking in place caste oppression in urban infrastructure (P. Gupta 2022; Prashad 2001). Urban local bodies were charged with collecting waste from municipal bins and transporting it to dumpsites. Municipal charters explicitly name solid waste as the state's property once it reaches the bin. This rule, though unevenly enforced, leaves waste pickers susceptible to harassment by police and street-level bureaucrats. In some cases, waste pickers were given access to dumpsites and transfer stations by individual functionaries of the state for diverse reasons, from kickbacks to identity-based relationships of solidarity. In other cases, municipalities permitted waste pickers to engage in door-to-door collection, granting access to waste in exchange for providing uncompensated labor.[15]

In the 1990s, a push to "clean and green" cities to make them more "world-class" complicated these arrangements. The problem of urban litter, as opposed to being linked to rapidly rising plastic use, was attributed to inadequate service provision from the informal economy. This prompted calls for municipalization and formalization of waste collection (Luthra 2019). With the passage of the 2000 Municipal Solid Waste Management Rules, municipalities across the country looked to bring door-to-door waste collection into their ambit. However, this drive for formalized, universal service provision was soon superseded by a neoliberal privatization agenda advocated by the World Bank/IMF, which forced Indian cities to privatize many basic municipal services (Schindler, Demaria, and Pandit 2012). In the early

2000s, the World Bank–sponsored Jawaharlal Nehru National Urban Renewal Mission propelled the municipality to outsource door-to-door collection and waste transport to private companies. Most of Bengaluru's garbage contractors, sometimes referred to as the city's "garbage mafia," belong to powerful land-owning castes. Many have close familial ties to elected city councilors and a strong grip on municipal infrastructures (Shree D.N. 2017).

Contractualization worsened the terms of work for street-sweepers, waste collectors, truck drivers, and transfer-station operators, who constitute the "formal" solid waste workforce of the city. The Bengaluru municipality implemented a freeze on hiring sanitation workers into permanent positions in the 1990s (Narayanareddy 2011). Whereas permanent sanitation workers had access to pensions and held secure government jobs with benefits, contract sanitation workers (*guttige pourakarmikas*) receive no benefits, cannot count on job security, and are rarely paid on time. Most sanitation workers in India hail from Scheduled Castes and are expected to carry out a range of additional tasks that expose them to toxicity and harm, such as the transport of cadavers or manual cleaning of sewers (Lee 2017; Prashad 2001; Sreenath 2019). Their oppressed-caste status makes it easier for garbage contractors to deny them regular pay and force them to manually substitute for modern disposal technologies.[16] Neoliberalizing processes, in this way, cohere with preexisting patterns of racialization to reproduce casted/raced inequalities (Pulido 2017; Ranganathan 2021).

Over the past decade, trade unions representing *guttige pourakarmikas* have engaged in several strikes to bring attention to these labor conditions. The risks and hazards *pourakarmikas* faced during the COVID-19 pandemic brought more attention to their precarious position, and some recognition as frontline public-health workers (*The News Minute* 2021; Swaroop and Lee 2021). Recently, *pourakarmikas* have won some significant victories including a direct-pay system for street-sweepers and, in 2022, the regularization of a subset of the state's garbage collectors (Kalappa 2022). While these victories are testament to the power of collective action by well-organized sanitation workers, the state continues to renege on many of its promises (*Express News Service* 2022).

For waste pickers, privatized door-to-door collection of waste further restricted access to recyclable materials by enclosing municipal waste and creating new property regimes. When private actors take over transfer stations and dumps, waste pickers can lose access to the spaces needed to both collect and sort waste (Chaturvedi and Gidwani 2011; Samson 2015). In Bengaluru, Delhi, and elsewhere, waste pickers were kicked off dumpsites when these were handed over to private contractors. Waste pickers now must bargain with salaried employees of private contractors for access to dumps, in addition to dealing with municipal field staff or beat cops who occasionally harass them or demand bribes. Casualized job arrangements, unpaid wages, and bonded labor mean that sanitation workers often sell the recyclables they glean from waste to augment their pay, causing resource conflict with informal waste pickers. These privatized regimes reduce access and tend to burden women waste pickers in particular (Chandran, Narayanan, and Subramanian 2019; Chaturvedi 2014; Wittmer 2022).

Alongside the privatization of waste management, the increased invocation of postconsumer solid waste as a resource results in a broader resourcification of waste.[17] Although municipal waste has long served as a commons for waste pickers, governments or corporations enclosing and extracting profit from the materials in waste is a more recent phenomenon (Gidwani 2013a). Starting in the 1970s, recycling programs in the West—promoted in many cases by the plastic industry itself—started drawing attention to "wealth in waste," portending a shift in how municipal waste was understood (MacBride 2011). The early 2010s brought a wider panic about the scarcity of precious metals, resulting in attempts "mine" these from e-waste.[18] As emphasis was placed on the resource value of postconsumer waste, supply chains were created to reclaim this value (O'Neill 2019). At global environmental and resource forums, incineration technologies, sometimes designated waste-to-energy, are touted as ways to recover value from waste by turning it into fuel. In popular culture, the numerous reality television shows about people making significant quantities of money by reclaiming and repurposing other people's trash, including Money from Nothing in the United Kingdom and Trash to Cash in the United States, also reflect this trend. Global circular economy research and policy, with its "closing the loop" rhetoric, are further intensifying waste resourcification (as I discuss in chapter 5).

Resourcification of waste is now a global phenomenon that has turned municipal solid waste into a resource and profit frontier. It seeks to secure the circuits and spaces through which waste travels, intensifying property claims by the state and private corporations. Today, the state and intergovernmental agencies and global capital interests have turned their eyes to urban waste as a commodity frontier or mine.[19] Resourcification has brought new claimants to the value in waste as it enclosed and marketized waste commons, threatening waste-picker access in the process. In India, resourcification is intensifying in a cultural context in which waste work is casted work, and caste functions as a normalized mechanism of inflicting humiliation, violence, and oppression (Guru 2011; Sreenath 2019). Thus, while waste might be seen as a source of value, and as wasted spaces become valuable spaces, waste pickers who have known of and extracted this value for the benefit of the city and its economies remain discarded.

These dynamics are not unique to India. The Barcelona Research Group on Informal Recyclers counts more than fifty conflicts worldwide, where the livelihoods of informal waste pickers and recyclers have been harmed by policy shifts promoting privatization (Cairo and Johannesburg), incineration (Delhi, Ethiopia, and the Philippines), or by spatial ordering schemes (Vietnam, Uruguay).[20] Waste pickers and their allies have responded by forming trade unions (Uruguay, Pune, Ghana), cooperatives (Brazil, Argentina, South Africa), and other types of organizations to defend their the right to waste. In the discussion that follows, I describe the case of Parisara Tanda, an organization representing waste pickers in Bengaluru, to argue that, in cities where waste is "sustainability terrain" and the playground of elite environmentalists, waste pickers deal with new pressures and scrutiny. Communal sustainability also opens new avenues for claims-making and political negotiation.

Relational Organizing for Cross-Class Solidarity

As middle-class interest in recycling intensified in Bengaluru, the voices of groups that advocated for the rights of informal waste workers began to insert themselves into policy conversations. One such group is Parisara Tanda, an organization working with waste pickers and small scrap dealers.[21] A longtime labor organizer who had previously worked with waste pickers in another Indian city launched Parisara Tanda in 2012. The fact that Parisara Tanda was initiated by non-waste pickers was itself not exceptional. Indeed, across the world, waste picker organizing has been largely initiated by middle-class actors.[22] However, the organizational forms that emerged in Bengaluru were constrained by the city's highly privatized waste management systems, the timing of the organizing effort, and the agendas of middle-class zero-waste activists.

Middle-class schemes to reform waste management in the city threatened waste-picker livelihoods. When key zero-waste organizations like the Clean and Green Forum (TCGF) were convened in 2009, they framed waste as primarily a public health and environmental issue, ignoring its livelihood dimensions. TCGF wanted to recruit private companies to collect recyclable "dry" waste directly from households. Such a scheme would cut out waste pickers and *kabadiwallahs* from waste flows. In my first interview with Parisara Tanda's founder, they explained that the organization was formed as a way for waste pickers to participate in city-level policy conversations:

> Waste pickers in Bengaluru are not organized. There are all these policy discussions happening which will affect them, but no representatives of waste pickers or scrap dealers are in the room. Yes, there are some local organizations operating in waste picker colonies, but this is not enough. I have seen all over the country that when middle class get involved in waste management, waste pickers get wiped out. . . . Well they may not be wiped out but access to waste gets much harder. They may have to travel further to get it.

A sense of urgency and a desire to take advantage of the opportunities that the 2012 public interest litigation (PIL) presented shaped Parisara Tanda's early tactics and strategies. I felt the urgency too, which is why I jumped in to assist the organization in its work, extending my fieldwork by six months. Everything seemed to be moving fast. The High Court judges, exasperated with the status quo, seemed to be willing to try new things. They issued orders every week, which, if implemented, would affect every aspect of the municipal solid waste system. As groups like TCGF searched for corporate saviors, inserting waste pickers into the conversation as part of the solution and not part of the problem was an urgent imperative. Needing help, Parisara Tanda turned to Women in Informal Employment: Globalizing and Organizing (WIEGO), an international organization that supports member-based organizations in the informal sector, for funding and advocacy support. Case studies of waste-picker integration were shared among the community, to communicate, "It has been done, we can do it here too."

The first task was organizing the city's waste pickers, who lived across several informal settlements. In 2012 and 2013, five nongovernmental organizations (NGOs), which made up Parisara Tanda's grassroots network at that time, began reaching out to waste pickers as individuals, families, and communities. Every Wednesday afternoon, twenty or thirty waste pickers came to the makeshift space provided to Parisara Tanda by a local community radio station for meetings. Most of the attendees were women. Five or six of them came with small children. We drank tea, ate biscuits, and discussed the situation in Bengaluru. The office had only a few chairs, so we sat on the floor, as one of the staff members, a longtime social activist in the city, explained how new rules mandating that all dry waste should go to materials recovery centers would threaten waste-picker livelihoods. He said in Tamil, "There will be less waste on the streets. More people want the waste now," illustrating the need to organize.

Parisara Tanda did not limit its outreach activities to free-roaming waste pickers. Rather, it looked to build relationships with informal itinerant buyers, scrap dealers, and aggregators in the city with the view that uniting these actors together under the same banner would build a stronger political coalition. It forsook the trade union mode of organizing. Instead, it argued that more effective outcomes could be achieved through building cross-occupation solidarity and working closely with middle-class actors. Consequently, it remains distant from the All India Central Council of Trade Unions (AICCTU) Karnataka, the Pourakarmika Sangha, and other trade unions representing Bengaluru's contract sanitation workers, who adopt a class-based organizing approach and make claims directly to the state.[23] Instead, Parisara Tanda pursued a strategy of relational organizing for interest convergence.

Recognizing the extent of middle class interference in Bengaluru's civic and environmental infrastructures, Parisara Tanda sought to build partnerships with elite and middle-class zero-waste movement (ZWM) leaders. During one of our early interactions, two Parisara Tanda staff defended this organizing approach: "I really feel that for any policy on waste to work, you have to partner with middle class. . . . If they do not cooperate, it won't work. Not everyone in the labor movement agreed but we thought, let us try to experiment here." This strategy of recruiting urban elite and middle-class support has been used by rural and livelihood-based environmental campaigns across India, because elite opinion has significant sway with the courts and political authorities (Kashwan 2023).

The class/caste orientation of Parisara Tanda also determined where and in what fora claims were made: Parisara Tanda made arguments for the inclusion of waste pickers at waste management and environmental festivals, on social media, and in the courts. Seeing that TCGF had already emerged as powerful gatekeepers with the ear of the state, Parisara Tanda decided to become a co-petitioner in the 2012 PIL in the Karnataka High Court. This organizing effort did not spill onto the street until several years later, in 2016, when Parisara Tanda–affiliated waste pickers and their allies protested unfair inclusion terms in front of Bengaluru's city hall. Until then, claims-making was largely done in a convivial mode. There were more social media posts than street protests or strikes in its repertoire. Bengaluru's penchant for elite forums and extraconstitutional urban planning and decision-making bodies, especially in the domain of waste management, also motivated the strategic engagement in these civil society spaces. But while Parisara Tanda collaborates with TCGF and other organizations that now support informal inclusion, it has more conflictual relationships with technocratic members of "Brand Bengaluru" (Gopakumar 2020) who question the capacity of informal workers to professionalize. Thus, in some spaces it goes along with middle-class opinions for strategic reasons, while in other spaces they push back to defend waste pickers' rights to waste.

Local complexities and the timing of events also meant that the organization's practices differed from how waste pickers have been organized or have organized themselves in other contexts.[24] Unlike a bottom-up process of organizing (often through trade unions and cooperatives) followed by claims-making for inclusion into municipal systems as door-to-door collectors (as in Pune and Delhi) or to operate materials recovery centers (as in Mumbai), claims for integration had to be made while waste pickers were still being organized in Bengaluru. One of the PIL demands submitted by Parisara Tanda in 2012 asked the BBMP to sponsor an effort to enumerate waste pickers. This demand was successful, and after being ordered to do so by the courts, the BBMP funded health camps to help Parisara Tanda register waste pickers into a database. This in turn helped those waste pickers obtain occupational identity cards that legitimized their occupations and offered some protection from police harassment. Registered waste pickers also gained access to health and education schemes designed for members of Schedule Castes and Tribes, but which they had not previously benefited from, due to lack of awareness or appropriate documentation. At the same time, because door-to-door waste collection was already being carried out by contract sanitation workers, there was no obvious formalized employment opportunity available for waste pickers to seek. This forced Parisara Tanda to pursue more entrepreneurial routes to retain waste pickers' access to waste.

The organization's leadership is composed of longtime labor, social justice, and education advocates recruited from Bengaluru's middle and lower-middle classes, as well as a few vocal and highly visible waste pickers who represent Parisara Tanda at public events. As of 2022, Parisara Tanda has over one hundred staff and works across eighteen cities and sixty-four villages. It is formally registered as a trust and operated by a professional staff, though there are several forums through which waste pickers participate in determining organizational strategies, such as weekly meetings in *bastis* [informal settlements] and monthly meetings at the organizations' offices in central Bengaluru. Nevertheless, only 11 percent of the organizations' staff hail from waste-picking communities and none are in senior leadership roles yet. While the organization initially aspired to function as a cooperative, it works more like an NGO, providing services to its constituent groups.[25] As of 2022, Parisara Tanda has provided services to more than twenty-seven thousand waste pickers across Karnataka state, helping them obtain occupational identity cards, scholarships for their children, and access to social service schemes. Most of its beneficiaries are women and it takes a gender-sensitive approach in its work. It has workstreams on livelihoods, policy, housing, health, social security, and education, all led by professional staff. Yet, it explicitly supports the rights of waste pickers to collectivize, self-represent, and engage in political advocacy directly: it has a civic leadership training program in which eight hundred waste pickers have participated, fifty of whom are now engaged in a special program focused on environmental law and rights.

Parisara Tanda experienced remarkable successes in a short period. Many prominent middle-class zero-waste advocates have gone from supporting the corporate takeover of recycling infrastructures to advocating for waste-picker inclusion in various legal and policy forums. As one ZWM leader remarked to me in February 2013,

> When we started, we came at this as a public health and environmental problem. The priority was to keep the waste off the streets and to manage it responsibly. We saw that the government and its subcontractors were not doing this, so we started looking to corporate actors, CSR, and some NGOs. We did not really know about the informal sector. Waste pickers were not on our radar. But now the issue of marginalization of waste pickers has come to the fore and now become an important part of our activities because of the creation of Parisara Tanda.

In the discussion that follows, I analyze the discursive frames developed by Parisara Tanda to talk about waste pickers and their work, and to make claims for waste access and for inclusion into emerging ZWM infrastructures. The claims-making discourses employed by Parisara Tanda were also occasioned by the increased invocation of environment in India's national waste management policy frameworks (Gill 2021). I argue that framing waste pickers as robust entrepreneurs and silent environmentalists created alliances with elite and middle-class zero-waste organizations that were primarily concerned about environmental and livability issues, and that this tactic of alliance-building was crucial to advancing waste-picker claims for belonging and inclusion in this hyper-neoliberal city.

Leveraging the Biases of Performative Environmentalism

Members of Bengaluru's waste-engaged middle classes, like performative environmentalists worldwide, see practices like segregating waste, recycling, and composting as virtuous and a source of moral authority. But as I argued in chapter 2, workers are denied this ecological legitimacy because of performative environmentalism's class and caste biases. Every member of the ZWM I interviewed displayed a well-informed commitment to ecosystem health and sustainability, but almost no one made the connection between waste management, waste work, caste, or class. It was as if the question of who labored to make waste manageable, under what conditions, and for what outcomes was a thorny subject only to be engaged with a managerial disposition.

Parisara Tanda's initial goal was to change these perceptions. Rehabilitating the public image of waste pickers and recyclers has been a goal of waste-picker groups across the world (Dias 2016; Chikarmane and Narayan 2005; Samson 2016). Labor organizers argue that constituting a collective dignified identity for waste pickers is a necessary first step before convening strategic partnerships (Kain et al. 2022; Samson 2009). With this explicit goal of contesting deficit-based understandings of waste pickers, the organization's mission statement links livelihoods and the environment, arguing that they seek to both improve the lives of waste pickers and, through this, improve how waste pickers help the city.[26]

Silent Environmentalists

"Parisara Tanda," the organization's name in Kannada chosen by waste pickers in an early meeting, loosely translates to "green force" or "green team" in English. Through its name, Parisara Tanda very consciously casts waste pickers as environmental subjects who work at the margins to provide a hitherto undervalued and misrecognized service to maintain city environments or, as a 2012 poster proclaimed, as "silent environmentalists" (see figure 3.1). This characterization is evident in the organization's signage and logo, in the green jackets that members wear, and in the ways in which waste pickers and their representatives present themselves at public forums and meetings. Presenting waste pickers primarily as environmental actors is a strategic and pragmatic tactic to build connections with environmental groups. As one Parisara Tanda organizer says, "We are not excluding any issues and saying its only waste pickers' concerns, right. We are also talking about waste, environment issues. . . . But we are coming as waste pickers to talk about the environment."

It is also in the context of the constitutive link between ecological legitimacy and the duties of citizenship implied in performative environmentalism that the use of the term team or force becomes interesting. Force invokes waste pickers as warriors for the collective good and even has some nationalist

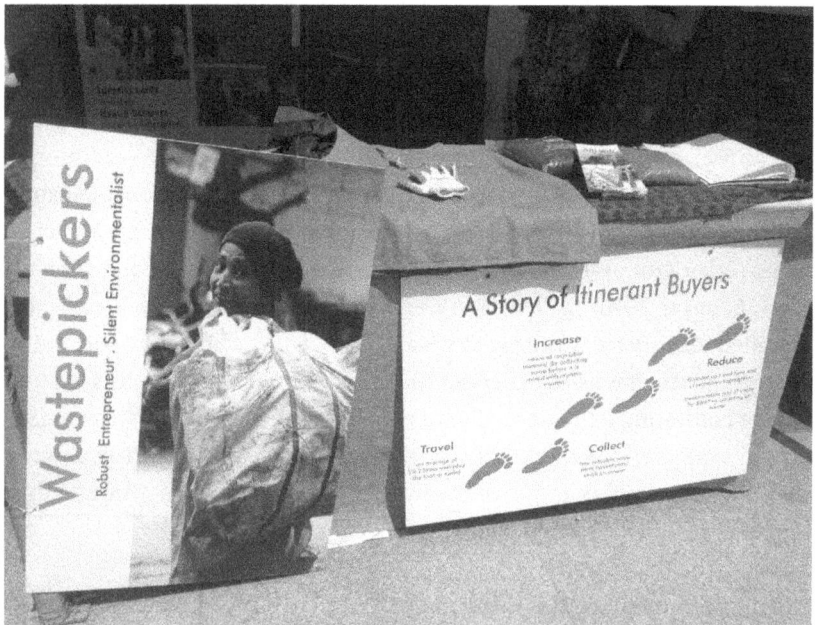

Figure 3.1
Parisara Tanda stall at "Wake Up Clean Up" event featuring the uniforms and gloves worn by their waste pickers.

overtones. This emphasis on duties and obligations versus rights and injustice calls up the image of a citizen who takes up arms in defense of the city-state or, in this context, risks their health and well-being to defend the city from the menace of garbage (see figure 3.2, which was produced as part of social media campaign that presented waste pickers as ecowarriors and "the Unwasters"[27]).

This narrative of highlighting labor and infrastructural contributions to make claims for citizenship is also part of the dynamic that Rosalind Fredericks (2018) describes in her study of waste infrastructure and citizenship in Dakar. Fredericks finds that trash workers personally and publicly frame their labor as an act of Muslim piety rooted in the spiritual value of cleanliness. This refusal of disposability turns the stigma of trash work on its head. The reframing of waste work away from a focus on livelihoods or poverty to a focus on public benefit enables waste pickers to dodge accusations that they are acting out of "vested interests," as other urban constituencies like street vendors have been accused of doing, or that they are the unruly poor

Figure 3.2
"The Unwasters" poster created by Parisara Tanda. This poster was shared in a tweet for World Environment Day in June 2019. Reprinted with permission.

soiling cities with their unhygienic citizenship (Anjaria 2009; Doron 2016). Parisara Tanda has been particularly successful at highlighting the environmental benefits produced from waste picking. In 2016, it received an award from the Karnataka government on World Environment Day—the first time a waste-picker organization was awarded an environmental award in India.

This strategy of utilizing environmental discourses has emerged in waste-picker claims-making across the world. For example, the NGO Chintan, which works with waste pickers in Delhi, published a report titled "Cooling Agents" to highlight the role that waste pickers play in mitigating greenhouse gas emissions (Chaturvedi 2009). Sociologist Manuel Rosaldo has chronicled how political artists in Bogota, Colombia, painted murals on the buggies used by reclaimers to communicate their status as invisible environmental heroes (Rosaldo and Alegre 2016). WIEGO has published several reports on the environmental contributions of waste pickers in cities and has sent waste-picker delegations to UN Climate Summits. Even powerful environmental organizations are lauding waste pickers' environmental contributions: UN Environment highlighted the environmental work of Pune's waste pickers as part of the 2018 World Environment Day festivities (UNEP 2018). A 2019 article on waste pickers in *Scientific American* proclaimed in its header, "A green army is ready to keep plastic waste out of the ocean. Unfortunately, the world is ignoring it" (Talbott 2019). Scholars like myself are also contributing to this framing, writing articles and research papers about the environmental contributions of waste pickers (M. Anantharaman and Schroeder 2021; Gutberlet and Carenzo 2020).

"We Are the Ones Who Will Touch the Waste"
"We are the ones who will touch the waste," declared Srima, one of Parisara Tanda's field managers at a meeting of apartment complex homeowners' associations organized in 2013. This statement, which was made as part of Srima's presentation on the waste management services provided by Parisara Tanda, emphasized how waste pickers are the people who willingly and dutifully put themselves in physical contact with waste, sorting it into multiple varieties and enabling the environmentally responsible regeneration of these materials. In invoking touch, it appeared to me that Srima was stealthily playing to the caste biases of her mostly elite and dominant-caste audience, who were bound to think of waste pickers as "untouchables"— even though they would likely not admit to this if confronted.

Srima was speaking at an event organized by Housing Habba, a company that sells data management and social networking software to exclusive, gated communities. This software tracks staff rosters, bills, budgets, and guest registries, all part of the disciplining tools of enclave urbanism. The audience was composed of homeowners, likely dominant-caste and upper-middle-class individuals who served on homeowners' association boards (a.k.a local busybodies). Srima was promoting Parisara Tanda's services and fighting off competition from other waste management service providers from the private sector, some of whom were touting technologies that they claimed could sort and bale waste quickly, with limited dependence on labor. That Srima chose this statement as her way of distinguishing waste pickers from these other service providers resonated with me for weeks as I tried to understand interactions between middle-class zero-waste advocates and waste pickers.

To understand why Srima emphasized that waste pickers alone would manually sort waste, a careful interrogation of both caste politics and the material nature of municipal solid waste is necessary. Casteism has made waste work an occupation imposed on Dalits and other oppressed castes because of the mythology that they are fated to do this stigmatized work because of their birth and identity (Teltumbde 2015). Caste oppression has also narrowed the already limited avenues for employment and economic advancement that Dalits have access to. While Narendra Modi's Swachh Bharat Mission has invoked the problematic Gandhian idea that all Indians must clean toilets or deal with waste as a way of destigmatizing the occupation, Dalit scholars have argued that this move has opened room for private players to takeover solid waste management functions while keeping Dalit workers in stigmatized roles (Teltumbde 2015). While private players hire (usually oppressed-caste) labor to manually collect, sort, and process waste, they simultaneously promote energy-intensive technology like street-sweeping robots or incineration to replace labor (Doron and Jeffrey 2018; Kornberg 2019b).

These technologies make recycling and resource recovery more difficult. For example, one company at the meeting recommended that each large complex invest in a baler that would compress all dry waste, reducing how much residents would have to pay for transportation and disposal. When one audience member asked whether this would make recycling more challenging, the speaker had no response. By contrast, communal sustainability proponents favored local, manual, and vernacular solutions, which have a metabolic dependence on labor. Parisara Tanda sought to leverage this dependence to demonstrate that recovering value from mixed municipal waste in a society of plastic proliferation requires touch.

The statement made by Srima positioned waste-picker work as an instance of virtuosic ability that enabled the alchemic transformation of waste into value, countering the narrative that their work is unorganized or unproductive. It is touch that classifies dry waste into twenty-seven categories. It is how metal lids are separated from glass bottles, how toothpaste caps are extracted painstakingly from discarded tubes. Waste pickers can distinguish multiple types of plastics based on sound, touch, and smell. Each category of scrap is then diverted to a different recycling stream, a productive afterlife. Focusing on touch makes legible how the work of picking and sorting is necessary for recovering value from waste and managing it in an environmentally responsible way.

It was also a challenge to their competitors, whose representatives could not bring themselves to touch the waste they claimed to manage with their technologies. Indeed, while representatives from start-ups and more established waste management companies made PowerPoint presentations to tout their wares, Srima and the two waste pickers at the event demonstrated the act of manual sorting (see figure 3.3). Using this visual tactic, Parisara Tanda presented a compelling argument that recycling and managing municipal solid waste required touch. It was a claim to expertise and utility while simultaneously an admonishment of the status quo in which waste pickers are devalued, corporate greenwashing is ubiquitous, and the elite look to outsource their dirty work to the cheapest bidder.

Presenting waste pickers as silent environmentalists and green warriors is an act of strategic essentialism that downplays the social (casted) aspects of waste-picking work and its connection to poverty, and instead focuses on the environmental benefits of the work. Waste pickers have always been essentialized in the popular discourse of Indian cities as dirty, dangerous, and antisocial. The strategy pursued by Parisara Tanda continues to essentialize waste pickers, but in a manner that *serves* their claims to waste and space. However, in doing so, Parisara Tanda also inadvertently reinforces the ontological connection between caste and waste enforced by Brahminism. Indeed, this is one of the reasons why traditional Ambedkarite organizations who argue that oppressed-caste people should not do waste work have not seen eye to eye with Parisara Tanda or with waste-picker organizations in other Indian cities.

Figure 3.3
Srima explains the benefits of manual sorting to two event attendees.

Discourses of Utility: Creating Robust Entrepreneurs

Much of the discourse used by Parisara Tanda in these elite-dominated forums referred to the contributions and services provided to the city by waste pickers. Even though "right to waste" is a major slogan of the Global Alliance of Wastepickers, I heard limited reference to this in their early organizing discourses. Instead, Parisara Tanda advocated for the inclusion of waste pickers into waste management systems by focusing not on waste-picker rights, but on their utility as entrepreneurs. This pragmatism emerged in the context of the state's gradual building up of a brazen record of rights violations targeting the urban poor, particularly those belonging to oppressed castes and religious minorities (Bhan 2016; P. Chacko 2018; Ranganathan 2021). Parisara Tanda also used these words because they believed they were more likely to get middle-class support for their activism if they emphasized what waste pickers could do, as opposed to what they were due.

These references to the economic utility of waste picker's work were frequently expressed in terms of the *lakhs* or *crores* of rupees that waste pickers

saved the municipality every day or month. An excerpt from a report published by members of TCGF and Parisara Tanda illustrates this emphasis:

> In 2013, we at Parisara Tanda along with KR University, Bengaluru, decided to
> look at the contribution of the waste-pickers of the city. The findings revealed
> that [registered waste pickers] save the city about twenty-three crores annually.
> Extrapolating the data for the estimated fifteen thousand waste-pickers that
> city harbours, the study revealed that the savings were about eighty-four crores
> annually.[28]

The focus on utility and value creation highlights the ways in which the
work of waste picking is productive and important beyond enabling the survival of waste pickers. It is a means of communicating the surplus value that
waste pickers generate by lowering the waste disposal burdens to municipalities in highly cost-effective ways.[29]

The act of refusing to see waste pickers as engaged in productive work
providing critical environmental and economic services is key to legitimiz-
ing their dispossession (Corwin 2020; Rosaldo 2016; Samson 2016). The
liberal polity's bias in seeing only those engaged in productive labor as
political humans with rights has been the basis of several decades of policies looking to eliminate or reform workers in the informal economy.[30] A
focus on value-creation serves as a counterstory to this invisibilization by
presenting waste pickers as vital agents enabling the productive life of cities. Demonstrating how they labor, through touch, to turn waste into value
enables further claims to political rights and eventually participation in
civil society. Attaching rupee amounts to their work makes it legible in a
neoliberal city obsessed with cost-effective service provision.

Neoliberal logics also structure narrative descriptions of waste-picker
labor: waste pickers are mobilized not as workers but as entrepreneurs. The
narrative of entrepreneurship also challenges the negative stereotype that
waste pickers are lone agents, instead highlighting positively their autonomy and independent streak.[31] Parisara Tanda thus took pains to repeatedly
clarify that they were not asking for waste pickers to be given jobs in the
formal waste collection sector, as I once witnessed during the PIL hearing in
April 2013. I recorded my observations in my research notes:

> It was a hot April afternoon in Bengaluru. The packed courtroom was steaming. At
> least 150 people were packed into the High Court chambers, where Justices Kumar
> and Nagarathna were listening to the public interest litigation Shankar vs. Govern-
> ment of India. On one side, the lawyer representing the petitioner and other NGOs,

including Parisara Tanda, who had impleaded themselves into the case. On the other, sat the hassled BBMP representative. Today, the matter being discussed was whether waste pickers should be recognized and given opportunities to participate in dry waste collection. Justice Kumar was raising objections to the note submitted by Parisara Tanda detailing the contributions of the informal sector and outlining arguments for their inclusion. "If you let them in, soon they will all be asking for government jobs," he said. The lawyer quickly replied, "No, no your honor, they won't. They don't need jobs[;] they will make money from the waste as always."

Government jobs were, for a long time, the most sought-after employment in the city. Ensuring job security, health care, and a pension, they represented the promise of a postcolonial developmentalist state that cared for its citizens. These jobs are still in high demand. A 2018 vacancy for fourteen street-sweeping jobs in Chennai attracted four thousand applicants, most of whom had college degrees (Rajan 2019). Openings for a hundred thousand jobs in the government-run Indian railways, where my father worked, attracted twenty million applicants (*BBC News* 2018). Those jobs, particularly in the municipal and state governments, have been slowly whittled away as part of neoliberal reforms imposed on cities and states by the national government and its suite of international advisers and creditors.

In a labor surplus economy where landlessness is on the rise, urban space is privatized, and salaried work is becoming more casualized, people coming to the city looking for work have limited options.[32] There are no jobs to give out, making the waged work-seeker persona non grata in the Indian city. But while a jobseeker is a pain in the state's side, an entrepreneur is a valorized figure. An entrepreneur is someone who makes something out of nothing or, in this case, value out of discards. The entrepreneur seeks no handouts, only investments that they promise to return. They do not ask for the state to care for them; they practice self-care and self-improvement. They take risks, experiment, and innovate. Bengaluru, which has nicknamed itself the start-up city, prides itself on creating an environment conducive to this type of experimentation and innovation.[33]

From Neoliberal Stigma to Entrepreneurial Valor

Countering the now-debunked claim that unwaged workers in the informal economy are unorganizable, informal workers are organizing effectively to defend their right to the city. They do this by asserting their identities as voters and organizers in their neighborhoods to influence electoral politics

(Agarwala 2013). They leverage constitutional statutes protecting democratic rights to make claims in courts (Rosaldo 2016). Knowing the fickleness of both the law and the political system, and because they are de facto denied the full rights of citizenship, they organize as communities along caste, religious, or occupational lines to make claims couched in the language of ethics to the state (Tucker and Devlin 2019). They maneuver social relations and social expectations to retain work opportunities, even when threatened by more powerful corporate actors (Butt 2019; Kornberg 2020). They utilize art and the tactic of culture jamming (Melo 2019; Rosaldo and Alegre 2016). They show up on the streets to display their numbers. They convene partnerships with groups with complementary and antagonistic interests (Kain et al. 2022; Zapata Campos et al. 2021). They educate members of the public about their work and demonstrate environmental stewardship (Gutberlet et al. 2021).

In Bengaluru, Parisara Tanda drew from this repertoire of organizing actions to stake their claims on behalf of waste pickers. However, the narratives and tactics of their claims-making were constrained by the governing logics of performative environmentalism and communal sustainability that valorize citizen-consumers and managerial expertise, while ignoring the role of working people in maintaining city environments. These sustainability logics, in combination with the postdevelopmental turn in how the poor are governed in Indian cities, have narrowed organizing options for waste pickers, while simultaneously opening new avenues for claims-making.

Engaging in a critical analysis of the material and discursive constraints imposed by the (neo)liberal and casteist logics of the city, and recognizing the ways dominant environmental discourses minoritize workers, Parisara Tanda sidestepped the language of work to instead characterize waste pickers as entrepreneurial environmentalists. Demonstrating to sustainability advocates that waste pickers do the actual work of segregation and recycling, organizers made visible these fundamental dependencies on labor, but without asking for the entitlements that waged workers might be due in the formal sector. They asked for a conducive regulatory environment and start-up capital, as opposed to rights or jobs for waste pickers. They spoke the languages of the neoliberal order as a way of seeking inclusion in the privatized and commodified city. Bengaluru's obsession with being seen as a start-up city and garden city made entrepreneurial environmentalism a compelling and effective category to mobilize for inclusion. A posture encompassed in the phrase *of use to the city*, a form of utility-based relational claims-making, has emerged as a dominant mode for Bengaluru's waste pickers to secure their rights to the city.

Yet this form of strategic essentialism, even if done in the name of justice, leaves still in place the structures of oppression and systems of power that create negative stereotypes of waste pickers (Liboiron and Lepawsky 2022, 112). Environmental activists came to support waste-picker claims because of how these responses enabled them to achieve their class- and caste-based interests. In the words of one Parisara Tanda staff member in 2017, "It has become cool for the middle classes to now say, I know a waste picker, that I have hired a waste picker for an event." This is an instance of pragmatic interest convergence between powerful and disenfranchised groups, but one that does not require the privileged to problematize the social structures that normalize stigma and gross inequality. It is also a strategy that leaves waste pickers vulnerable—if the middle classes or the state were to find a *more* useful group, nothing would stop them from switching loyalties—as these groups have made no ethical commitments to waste pickers. Further, the focus on individual entrepreneurship deemphasizes collective identity and organization, which could make it harder for waste pickers to mobilize the collective strength needed to halt policy changes that harm their prospects.

Waste-picker organizations across India have successfully fought for and gained recognition of waste picking. Waste pickers in several Indian cities are now formally recognized through occupational ID cards, effectively decriminalizing waste picking. Provisions protecting waste-picking livelihoods were included in the revised Solid Waste Management Rules issued in 2018 by the national government. The tide is shifting, and not just in India. In cities across Asia, Africa, and Latin America, member-based organizations and NGOs have successfully protested threats to livelihood from legalistic formalization and privatization projects, emphasizing the right to waste and livelihood for informal reclaimers (Demaria and Todt 2020). As waste pickers become more widely recognized for the economic and environmental value they produce, what sorts of inclusion are resulting from this recognition? The subsequent chapters explore the material arrangements and justice outcomes of this inclusion agenda in Bengaluru and beyond.

4   DIY Infrastructures

The municipal commissioner was in the hot seat. Justice Kumar, who was presiding over the public interest litigation (PIL) on solid waste management, had just roundly admonished him for not implementing court orders to set up materials recovery centers in each of Bengaluru's 198 wards. "Is this a joke to you?" Kumar bellowed, while the commissioner looked down at the ground. Under pressure to demonstrate good faith and compliance, the commissioner made an announcement: "The Bruhat Bengaluru Mahanagare Pallike [BBMP] assures the court that 198 dry waste collection centers will be operational within three months, and these dry waste collection centers will be the cornerstones for the decentralized management of dry waste in the city."

Bengaluru's zero-waste movement (ZWM) was also under pressure. It was desperate to convince naysayers that decentralized waste management infrastructures were feasible and scalable. Every time Justice Kumar wavered, the Clean and Green Forum's (TCGF) counsel made repeated assurances that their proposals were suitable for a city of Bengaluru's size and complexity. They knew that working models needed to be set up immediately and evidence of their success submitted to the court. The justices had been thus far supportive of their ideas, but things could shift quickly, as they had in the past.

Meanwhile, the BBMP's representatives were making subtle suggestions that Bengaluru could follow the lead of other Indian cities like Delhi and explore waste incineration instead. This was unacceptable to communal sustainability proponents, as the incinerator embodied the risk of ecological destruction. It also threatened to render their movement defunct. Conversely, for the state, waste-to-energy solutions represented the promise of spectacular infrastructure—the elusive but seductive silver bullet solution to the pesky problem of garbage. The smokestacks of incinerators were an invisible but potent presence in the courtroom.

Given this urgency, activists pushed the BBMP to commission a dry waste collection center (DWCC) in Freedom Park during the Wake Up Clean Up expo. A series of backdoor negotiations among the BBMP, Parisara Tanda, and middle-class mediators resulted in an unlikely outcome: the BBMP tasked Parisara Tanda waste pickers with conducting this DWCC demonstration. Within days, Parisara Tanda was able to get the center, which was nothing more than an empty shed, up and running. Periamma and her team worked fast, collecting recyclables from dustbins and transporting them to the shed, where three waste pickers worked to sort through the paper food containers, food wrappers, PET bottles, and aluminum cans. Parisara Tanda worked with students from a well-regarded design college to paint a mural on the shed, displaying zero-waste and planetary stewardship principles front and center (*Deccan Herald* 2013). The ecological credentials of the space would now be evident to curious passersby who might wonder about the purpose of this nondescript shed.

The commissioner finally inaugurated the DWCC on day five, with television cameras present. It looked functional and productive. Thanks to Periamma and her team, the BBMP was able to demonstrate compliance with court orders. Zero-waste proponents had something to point to while talking to journalists (see figure 4.1). The demo also helped display the speed, knowledge, flexibility, and hustle of waste pickers, who had turned an empty shed into a waste processing center with almost no equipment or resources. While endless panel discussions raged on with no resolution in the freezing air-conditioned auditorium, the real action was outside, where waste pickers affiliated with Parisara Tanda were resolutely demonstrating that they could deliver the clean and green cities desired by elites through this form of DIY infrastructuring.

What sorts of infrastructural arrangements emerge out of communal sustainability and entrepreneurial environmentalisms in Bengaluru? How did Bengaluru, a city with aspirations to become India's Silicon Valley, adopt decidedly low-tech infrastructure to solve one of its biggest urban challenges? The answer, I argue, lies in the way that environmental subjectivities and imaginations of elites rub up against the material and political-economic metabolisms of the city, mediated by the work and knowledge claims of informal waste pickers. It also depends on how local governments

Figure 4.1
Inauguration of Freedom Park Dry Waste Collection Center. In the foreground, a zero-waste leader from TCGF talks to TV reporters about dry waste collection. In the background, two waste pickers sort through the dry waste they reclaimed at the event.

navigate environmental claims made by different groups while looking to appease the demands placed on it by national programs that are now ranking cities based on cleanliness and, by extension, profitability.

In this chapter, I trace the history and evolution of the dry waste collection center concept, a coproduced DIY infrastructural arrangement that is both a product and vehicle for cross-class collective action. Against assumptions that technical and policy elites dominate sustainability in neoliberal cities (Hodson and Marvin 2014), zero-waste infrastructuring in Bengaluru was both democratic and democratizing. DWCCs enclose civic activism by middle-class women and entrepreneurial hustle by waste pickers to manage waste flows in the city. They also enable the inclusion and participation of diverse urban citizens in redesigning public services. Yet, DIY infrastructures are contradictory, exposing waste pickers to new risks and responsibility while, at the same time, advancing more democratized and participatory forms of communal sustainability.

Infrastructuring Sustainability

City governments across the world are creating more infrastructures in the name of sustainability and decarbonization.[1] These infrastructural undertakings range from demarcating bike lanes on busy streets to placate bicycle advocates and project economic vitality in urban cores, to constructing massive sea walls to protect the valuable seafront homes of the 1 percent. Making (sustainability) infrastructure is a contentious process. Engineers, planners, and city managers are key protagonists tasked with reconciling sustainability goals with budgetary constraints, corporate lobbies, quality of life concerns, and increasingly, questions of equity. Making infrastructures, or infrastructuring, is no longer the exclusive purview of the state.[2] Rather, the push toward privatization, which manifests neoliberal logics of economic efficiency, municipal divestment, and de-risking, has meant that infrastructures are created and governed by a hodgepodge of public and private actors.[3] In schemes such as New York's Citi Bike program, cash-strapped municipal governments invite corporate sponsors to collaborate in delivering services, offering lucrative branding and licensing opportunities in exchange for capital (Stehlin 2019). In African and Asian cities, communities band together to provision their own basic needs for water and sanitation (Mitlin and Bartlett 2018).

Evaluating the equity impacts of infrastructure development is a key preoccupation in critical urban studies.[4] In Asian cities, scholars have revealed how infrastructural projects are used to legitimize the destruction of informal settlements and economies in the name of public interest and the environmental good.[5] Infrastructure privatization usually worsens service provision for poorer neighborhoods, disproportionately harming women and racialized groups (Ranganathan 2016; Sultana 2018; Truelove 2019). Sustainability projects have led to green gentrification, a phenomenon in which urban greening, ostensibly carried out for the well-being of current residents (often people of color and working class), results in their displacement (Gould and Lewis 2016; Rice et al. 2020).

These inequitable outcomes are also precipitated by the aesthetic regimes of sustainability. Analyzing the mobilization of ideologies in urban sustainability thought, sociologists Hilary Angelo and David Waschmuth (2018) argue that nature-based "green" and high-modernist "gray" ideologies animate the aesthetic forms that urban sustainability manifests across the globe.[6]

Social technologies that do not mimic these registers are not readily recognized as sustainability. Yet, as we will see, dry waste collection centers do not easily map onto the rubrics of elite, modernist green-gray visions. While the DWCC is emblematic of a broader turn toward localized, public-private infrastructure for sustainable waste management (see Doron and Jeffrey 2018), it challenges what sustainability infrastructures can look like. In most cases, these centers appear neither green nor high-tech. Some of them have straggly potted plants at their entrances, but most are surrounded by bales of discards. A select few now have conveyor belts, balers, and other enhancements, but most lack these technological trappings. Examining the processes through which DWCCs emerged and evolved, I explore how high-tech Bengaluru ended up with a decidedly shabby infrastructure that makes room for informal waste pickers.[7]

Understanding infrastructuring as a process demands ethnographic attention and long-term engagement. It requires documenting the material and political lives of infrastructure to reveal the tensions between ordered plans and disordered results. This is where a situated urban political ecology (SUPE) approach becomes useful.[8] While infrastructure is commonly understood as physical or inanimate provisioning systems, comprising roads, pipes, and wires, urban political ecologists have drawn attention to how people, through practices and social relations, function as social infrastructures (Furlong 2011; Silver 2014). This approach sees infrastructures as "heterogeneous configurations" composed of people (Simone 2004) and of knowledge, productive, consumptive, and affective practices.[9] SUPE calls for ethnographic attention to the everydayness of infrastructure planning, performance, and dissimulation to characterize the social and political lives of networked systems of social reproduction in cities.[10] Such an approach is especially important because many sustainability infrastructures require involvement by ordinary people as citizens and consumers, who are expected to change their behaviors to help them function. After all, bike lanes only work if cars do not park on them (see figure 4.2).

Infrastructures both constitute and are constituted by diverse publics, who, in negotiation with the state and with each other, articulate visions of what is right, what is due, and what is enough. Infrastructure is a means through which excluded publics can make themselves known, seen, and heard. Tactical modes of infrastructural citizenship in Global South cities have ranged from paying for pipes to garbage strikes.[11] Attention to

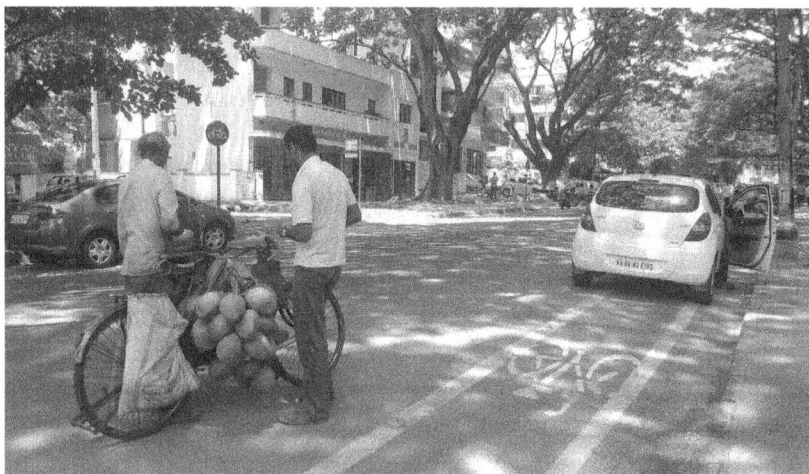

Figure 4.2
A car parked on a bicycle lane in the neighborhood of Jayanagar. In the foreground, you can see a working-class cyclist conducting his livelihood of selling tender coconuts.

infrastructure is also attention to sociality and the ways in which people come together, act on each other (and on things), and expand their own capacities (Simone 2004). In this vein, coproduction of essential public services like sanitation and waste management has been documented to be one route through which communities develop capacities, skills, and collective will and build relationships with various branches of the state (Mitlin 2008; Watson 2014). Yet, studies of infrastructural coproduction tend to skim over history, identity, and culture and flatten communities (Castán Broto and Neves Alves 2018; Unnikrishnan, Castán Broto, and Nagendra 2020). They often fail to identify how social location determines the ways in which different actors engage with an infrastructural arrangement (e.g., as consumers versus laborers), tending instead to lump diverse social groups into static categories.

In Bengaluru, the study of the DWCC as a DIY infrastructure bringing together the municipality, middle-class environmentalists, and waste pickers offers insight into the promise and limits of communal solutions to urban environmental problems. It demonstrates that the coproduction of waste infrastructures is formed around existing metabolic divisions of labor that are gendered, classed, and caste-based, highlighting the dynamics of care work and social reproduction in studies of infrastructural coproduction. I suggest that, for waste pickers, coproduced infrastructures can be simultaneously inclusive and exploitative. Ultimately, the degree to which DIY infrastructuring realizes equitable and sustainable cities is contingent on whether coproduction shifts power relations, in addition to providing services or livelihood opportunities.

Tracing the Dry Waste Collection Center

A dry waste collection center (DWCC), in its essence, is a materials recovery facility. Its functions are simple in theory: it serves as a flexible space where dry waste is managed and converted to value, both economic and environmental. Economic value is realized by the tasks of storing and categorizing dry waste into saleable categories, turning waste into a tradeable commodity. Environmental value accrues by the avoidance of waste on the streets, protecting middle-class eyes from aesthetic blight, cows from consuming indigestible materials, and waterways from polluting microplastics. Like any waste infrastructure, it is a technology designed to redistribute waste's burden (Stamatopoulou-Robbins 2019, 24)—but also, in this case, to exploit the value in discarded materials.

My first introduction to the dry waste collection center concept came at a TCGF meeting I attended in 2012. A forum member was presenting a draft proposal that was to be submitted to the High Court. My initial impression of the DWCC, from the two-dimensional mock-up rendered on a PowerPoint slide, was that it was dimensionless: a space of alchemy where the transformation of waste to value was instant, thus rendering the requirements of physical space or structure rather immaterial. Named Kartavya, which means "our duty," the DWCC gave physical embodiment to moral and philosophical principles of communal sustainability—waste was a shared and collective responsibility to be borne by all urban residents doing their bit. Its proponents, members of Bengaluru's middle-class zerowaste movement, framed the DWCC as the desirable opposite of another popular infrastructural solution to the problem of garbage, the waste-to-energy, or incineration, plant. Incinerators are a prime example of the centralized, capital-intensive, and spectacular infrastructures that are heralded as symbols of development and modernity in majority world cities

(Schwenkel 2015). In contrast, the DWCC is decentralized, labor-intensive, flexible, and accessible.

In 2012–2013, the BBMP began the slow process of commissioning new centers and incorporating existing recycling facilities into its network. I observed several centers opened in a hurry by the BBMP to demonstrate compliance with High Court orders and to placate vocal activists who wanted to see their visions realized. In subsequent years, both English-language and vernacular media discussed dry waste collection centers in competing terms. Media reporting often praised DWCCs as models of sustainability and community participation, of salvation finally delivered to the garbage city. Other assessments described the DWCC as an unfulfilled promise, invoking images of decrepit sheds that look like mini dumps rather than productive sites of sustainable waste management. Every few months, press releases or reports describe how the BBMP is doing something to make DWCCs more efficient, such as fining people who do not segregate their waste or promising to rebuild damaged centers.

In the 2020s, DWCCs are still being built, relocated, and demolished. As of June 2022, Bengaluru had somewhere between 166 and 180 dry waste collection centers.[12] Exact numbers are hard to pin down. They remain a diverse bunch in their physical arrangements and material forms. Many are no more than a shed with a tin roof and four walls. One is a repurposed public toilet! Some have equipment in the form of conveyor belts and balers, but most DWCCs, especially those operated by waste pickers and scrap dealers, are in disrepair. Their current physical forms stand in stark contrast to initial mock-ups of what these centers should look like, which conjured images of recycling centers in Europe for inspiration.

In the discussion that follows, I draw on ethnographic fieldwork and review of key policy documents to identify three key factors that drove Bengaluru to DIY infrastructuring. By tracing the DWCC over a decade, I show how diverse actors negotiate each other's differential understandings of both problem and solution, incrementally changing the material and social relations of this heterogeneous infrastructural configuration.[13]

The Hazards of Incineration

To understand the emergence of the DWCC in Bengaluru, we have to look to other places. The first of these places is New Delhi, India, which in 2012 was the site of a bitter and much-publicized battle. On one side were the residents of Okhla and Ghazipur, alongside Delhi's waste pickers; on the other side were the central government and private companies who operated the city's two waste-to-energy (WTE) plants.[14] Delhi, in its persistent search for cost-efficient and privatized solutions to the problem of waste, had utilized Kyoto-protocol Clean Development Mechanism funds to construct WTE plants in the urban periphery. Indeed, incinerators were coming up all over the Global South as roving capital sought infrastructural solutions that could generate "clean energy" by removing dirty garbage, while also producing emissions credits for polluters in the EU and elsewhere.[15] WTE appealed to municipal and elite desires to spirit away garbage from the world-class city. It represented the infrastructural intensity that was considered (though mistakenly) easier to implement and manage—centralized control with corporate management.[16] WTE advocates argued that, given ever-rising waste generation rates in the city and the lack of other scalable solutions—choosing here to ignore or minimize the recycling work done by the informal sector—incineration was an urgent public health and environmental necessity.

Not everyone was convinced. Instead of laudatory reports about the effectiveness of WTE, troubling news about the Delhi situation reached zero-waste management advocates in Bengaluru via newspaper articles, Facebook, and WhatsApp: waste-to-energy plants were terrible for human health. The children of Okhla and Ghazipur could not breathe. Health problems were on the rise. Toxic ash was everywhere. Indian municipal waste had too much organic content; it was simply too wet to burn well. To feed the beast the high calorific waste that it needed to function, recyclables needed to be kept in the mix. Any attempt to remove paper and plastics, which were until then extracted by the informal sector and diverted to recycling streams, was halted immediately so that the waste-to-energy plant could function at maximum efficiency. Recycling was rendered undesirable by the material requirements of incineration, which in turn was in service of generating carbon offsets for Global North polluters.

For Bengaluru's communal sustainability proponents, this was a horrific proposition. Their mantra was that dry waste generation in the home should be minimized and, if generated, dry waste should be washed with care, sorted, and sent on to a productive afterlife of reuse and recycling. Discourses of environmental responsibility, *kartavya* (duty), and self-empowerment were central to their formation as environmental subjects. Physical contact with dry or even organic waste did not disgust Bengaluru's mostly dominant-caste and female zero-waste practitioners. One might expect it to, given their caste backgrounds and dispositions. But the purification agenda (Kornberg 2019b) that motivated waste to energy as a preferred solution in Delhi was not as important to this community because they recast embodied disgust at handling waste into a source of ecological pride and moral worth—something that is possible when you already have caste privilege.

Bengaluru's zero-waste proponents were also genuinely fearful of the health impacts of waste to energy. Interactions with organizations like the Global Alliance for Incinerator Alternatives (GAIA), a global network advocating against incineration and for zero waste, intensified their fears. Through GAIA, and with some googling, ZWM leaders in Bengaluru became proficient in articulating their opposition to WTE on public health grounds. They learned to deploy the language of toxicology and health to describe the risks of waste incineration. For example, a slide deck by a TCGF member that was circulated on Facebook and several email lists declared, "Plant will spew toxic gases into the surrounding air, and toxic ash will have to be disposed of. There are highly toxic residual gases like dioxins even with scrubbers and filters."

At waste management advocacy presentations, WTE was maligned as toxic and even sinister. These presentations, given mostly by middle-aged and older women, usually included a mention of the dangers that dioxins posed to reproductive health. The presenter would scan the audience looking at young mothers or would-be mothers and target their comments at them: "Dioxins will affect unborn and newborn babies the most. The impacts are long-term." To the men, they would direct rhetorical questions: "If Indian waste does not have high enough calorific value for incineration, then why is the government pushing this so aggressively as the solution to the garbage problem?" Implicit in their question was the speculation that corruption was part of the story. ZWM leaders were both convinced and convincing that waste to energy was part of a plot to make money at the expense of their health. Their arguments leveraged existing narratives of corruption and played on middle-class distrust of the state.[17] Waste to energy, while spectacular, was distant, opaque, and therefore threatening.

Embodied fear of environmental harm produces environmental subjects and motivates social action. In Delhi, middle-class communities living next to WTE plants formed an unlikely coalition with waste-picker unions trying to defend their livelihoods. This novel and unprecedented coalition, which expressed itself through *dharnas* (strikes), protest marches, and media advocacy, was forged through the shared experience of an existential threat—to health and well-being for one group, to livelihood for the other (Demaria and Schindler 2016). Political ecologist Sapana Doshi reminds us of the importance of paying attention to embodied experiences in understanding urban environmental politics, arguing that "embodied experiences and discursive logics of urban environmental rule contour political ecological claims" (Doshi 2019, 114). Environmental imaginations and policy mobilities matter as well. Although there was no incinerator in sight in Bengaluru, the very threat of it was visceral enough to invigorate staunch opposition. While it was a shared and embodied experience of pollution and precarity that brought middle-class associations and waste-picker groups together in Delhi, the imagined pollution and precarity posed by the very threat of waste-to-energy plants also convened an unlikely coalition of resistance in Bengaluru. The strength of the coalition was bolstered by the optimism that alternatives to incineration were possible. I next direct attention to the circuits of information and policy mobility that gave shape and form to the search for an infrastructural alternative to the smokestacks of incineration.

The Allure of Colorful Bins

The second set of places we must look to understand the shape, form, and logic of the DWCC includes spaces of transnational advocacy on zero waste, as well as cities that are globally celebrated as pioneers in implementing zero-waste principles. Organizations like GAIA, which oppose the burning of waste on environmental justice and ecological grounds, work in India and other Global South countries to promote community-driven waste management solutions as an alternative to incineration. In India, GAIA has an email listserv that is used by zero-waste activists from across the country as well as members of GAIA's teams in the United States, Europe, and the Philippines to share information. Participation in these lists helped some zero-waste advocates develop more ecologically oriented critiques of WTE plants, as the final slide in a highly circulated slide deck comparing waste to energy to zero-waste practices describes:

> The waste disposal issue is not about making Bengaluru look beautiful for visitors and investors by ruining someone else's backyard. Only a small mind will think this way. It is about fixing the issue with a holistic solution that does not ruin

anyone's backyard—not Bengaluru, not Mavallipura, not Mandur. A WTE [waste to energy] plant will have to be located somewhere, and it is going to pollute someone's backyard with toxic gases and residual ash.

GAIA's educational materials helped ZWM actors understand that WTE was not really a "solution" to unmanaged municipal waste, but rather a process of spatial displacement that would move the burden of the waste from within the city to its outskirts, as was already happening in Mandur and Mavallipura. This again went against the communal sustainability ethic to minimize pollution. Instead, GAIA warned that WTE would "lock-in" dependence on high-calorie plastic waste, suppressing policies aimed at waste reduction or recycling, as it had in Sweden (Corvellec, Zapata Campos, and Zapata 2013).

In their search for alternatives that were coherent with Bengaluru's desired "clean and green garden city" image, ZWM advocates also looked to models in other global cities. In the numerous community advocacy presentations that I attended, San Francisco was highlighted as a model to emulate because of its "clean waste recycling centers, progressive policies and conscientious implementation of zero-waste principles."[18] ZWM practitioners saw San Francisco as a model city because of the perception that it managed waste in a manner that delivered both cleanliness and ecological health.[19] Given Bengaluru's aspirations to become "India's Silicon Valley," the vision of San Francisco was all the more appealing. In addition, some ZWM leaders had visited San Francisco as part of their work in the IT sector, and their testimonies echoed what people heard through their social networks and the Internet. A slide deck prepared by a ZWM member included impressive statistics: "San Francisco. Has a Zero waste program. 78% waste is reused or recycled. Target 90% by 2020. Collection done by private waste management firm."

In another early mock-up by a TCGF activist, in May 2011, she described the DWCC as modeled on waste recycling centers in Germany and Sweden, with colorful bins and gleaming conveyor belts. To emphasize her point, she displayed some googled images of recycling centers in the two countries in which all waste was out of sight, carefully stored in containers. We saw families in cars coming to drop off their recyclables in a building surrounded by trees, communicating an almost idyllic scene of community harmony in urban nature. Even though these images have little to do with the reality of waste disposal in Indian cities, where a maid or *pourakarmika* is more likely to transport waste rather than a middle-class person in a car, these images reassured activists that, even though they were saying no to waste to energy, they would still be able to secure city spaces that matched performative environmentalists' aesthetic visions of a verdant garden city.

But here, too, caste and class prejudice played a role in shaping communal sustainability's infrastructural ideals. TCGF could have looked closer to home for models—many Bengaluru neighborhoods have scrap shops that buy high-value waste from households, waste pickers, and sanitation workers, amassing larger volumes and then reselling those materials to wholesalers and recyclers within and beyond the city (Gill 2009). Scrap shops are the indigenous material recovery facilities of Indian cities. But the ecological credentials of scrap dealers were suspect, as they bought only dry waste that had resale value. The ZWM's imported categories of dry and wet waste do not map cleanly onto the vernacular separation of *kooda* and *kabad*, the former referring to valueless waste and the latter to scrap. Scrap dealers were held responsible for plastic bags and food packaging that ended up on the street. In their defense, scrap dealers were only following the market logic of the resourcification project, as its true pioneers: by refusing to collect low or no-value waste, *kabadiwallahs* were in essence refusing to provide the hidden subsidy of carrying away valueless waste. It is what makes their business viable and profitable.

Instead, DWCCs played into middle-class environmentalist's sense of self-importance: entrusted with the safekeeping of the city, they are above vested interests and can uniquely vision and enact ethical environmentalism. Localism and decentralization are common themes in middle-class civic politics.[20] Further, the DWCC—with an infrastructural arrangement mandating behavior change at home, a full-fledged public sensitization campaign, the appointment of volunteer supervisors in neighborhoods, and the construction of highly visible infrastructure throughout the city—enables the performance of ecological citizenship by the elite and the middle classes, while simultaneously empowering them to monitor these infrastructural arrangements. Thus, even if certain waste-to-energy technologies might be more climate friendly according to some metrics, they failed to meet the ZWM's aesthetic and performative criteria for sustainability.

ZWM leaders thus imagined the DWCC as a communal, coproduced alternative to the large-scale spectacular infrastructure of WTE proposed by national and state governments in India. In contrast to an incineration plant (a city of Bengaluru's size would need two), the DWCC is localized and decentralized, with one planned for each of Bengaluru's 198 wards. Unlike a waste-to-energy plant, which is a fixed asset that would require years to build and several more to break even, the DWCC is flexible. It can be built in a matter of weeks, its location can be moved if necessary, and installation costs are in the range of 10 to 25 *lakhs* compared to 200 *crores* to build a waste-to-energy plant with 1,000 tons of capacity.[21] The DWCC is defiantly low-tech and low-cost, almost artisanal, promising more efficient delivery of services with reproductive labor contributions replacing limited state resources.[22] If the waste-to-energy plant promised the purification of waste by turning it into energy, the DWCC promised the extension of the productive life of these materials in their current form. Finally, unlike incineration, the promise of the DWCC was that it would not endanger human health via the production and dissemination of dangerous toxins. Instead, it would provide health and well-being to both city residents and communities living near dumpsites.

Selling Sustainable

Having conceptualized the dry waste collection center, the ZWM faced an uphill battle compelling the municipality to get on board. In Bengaluru, community-based environmental projects, such as those concerned with lake restoration, have had mixed success in gaining the BBMP's support (Nagendra and Ostrom 2014). One reason for this is that local governments face a lot of top-down pressure from national and state governments and from corporate elites. Conversely, they are less responsive or accountable to local communities or marginalized groups. Consequently, political economic considerations usually trump environmental or even quality of life concerns. This is the reason that, even though the Indian government had issued rules requiring municipalities to enforce source segregation of waste for better management as early as 2000, next to no efforts had been made toward implementation. Rather, the only part of municipal reforms that were enthusiastically pursued were provisions encouraging the privatization of waste collection and transport, mostly because these dovetailed with the neoliberal political-economic agenda (Luthra 2019). Recognizing these constraints, the ZWM moved to leverage neoliberal commonsense ideas of zero-subsidy service provision, revenue-neutrality, and financial sustainability to promote DWCCs.

Alongside aesthetics, the neoliberal common sense that infrastructural design should be revenue neutral and financially sustainable disciplines what counts as a successful sustainability intervention (Bulkeley et al. 2010; Hodson and Marvin 2014). Facing neoliberal austerity measures, municipal governments act as cost-saving business actors who either reduce costs by outsourcing public services or find ways to move those costs onto consumers through privatization (Bigger et al. 2018; Goldman 2011; Peck 2012). For the BBMP, managing Bengaluru's waste had long been a costly affair, attracting criticism both from higher-ups in the governments and from citizens. In 2011, the municipal authority started charging a supplementary solid waste management (SWM) fee as part of property taxes. Yet, it was failing to clean up the city of its garbage woes. SWM already accounted for 15–20 percent of the municipal yearly budget. The BBMP did not seem inclined to spend heavily on decentralized systems.

Recognizing this, TCGF argued forcefully that their decentralized waste management infrastructures would operate using sound business principles and encourage private-sector and citizen participation to reduce costs. To sell the DWCC to a city government looking to demonstrate its smart and green credentials, proponents emphasized that this scheme would require low capital investment and not produce costly externalities like toxic ash. Another part of the seduction of the DWCC for state elites in the early 2010s was that it broke with the old, centralized forms of operating the city to decentralize urban forms and enable the exercise of civic citizenship.

A zero-subsidy model of service provision permits the state to engage in de-risking by moving financial burdens onto nonstate actors like informal workers, NGOs, and households (see also Theodore 2020 in North American context). TCGF crunched numbers to show that the BBMP could recover its capital investment in five to ten years. Operating costs would be recouped by reclaiming value from waste, optimized through economies of scale. Economic viability would be ensured through "market-driven delivery mechanisms." If Bengaluru was going to take a chance on zero-waste management, the risk would not be borne by the state. According to a submission made in the High Court in 2016, the municipality spent an estimated 21 *crore* rupees to set up the physical infrastructure for 150 DWCCs, which, compared to the 580 *crore* rupees it pays every year to garbage contractors, is an extremely low sum.

In keeping with this marketized mindset, the first nominees for DWCC stewards were from the corporate sector. One initial model favored by TCGF promoted an important role for ITC, which operated a corporate social responsibility scheme called Wealth out of Waste (WoW). ITC was an attractive partner because ZWM members believed that industry participation would help regularize recycling work, bringing professionalism, efficiency, and better environmental conditions, in addition to providing financial investment and marketing support.[23] The preference for ITC reflects an uncritical preference for formal enterprises compared to informal work and enterprises, what geographer Jennifer Tucker calls the formalization fetish (Tucker and Anantharaman 2020). Formalization schemes can disrupt informal livelihoods because informal workers and enterprises cannot readily comply with tax, labor, or environmental regulations without additional investment or funding. It opens the door for privatization, as new entrants have opportunities to take over activities in the informal economy.

Unsurprisingly, initial sketches of decentralized recycling systems did not consider how these schemes would affect waste pickers, scrap dealers, itinerant buyers, and other members of the informal waste economy. For many ZWM members, these actors were no different from any other poor person. In an early interview with the founder of an environmental NGO, I asked her whether she collaborated with waste pickers in any of her projects. She defended herself from the implicit criticism in my question by saying, "No, but I employ other poor people. They are also worthy of getting jobs, right?" A representative from a new waste management start-up was less concerned about optics, saying, "Waste pickers are often drunk. They are also lone agents who do not like taking orders. It will be hard to make them work well. It is easier for us to find other people who need jobs and employ them." This statement echoes what Melanie Samson observes in waste-picker integration efforts in Johannesburg. Samson finds that the state relegates reclaimers to equal status with unemployed community members who have never worked with recyclables, thereby dismissing their experience and knowledge (Samson 2015, 2019). For Samson, these assumptions emerge from coloniality. In Bengaluru, caste compounds and amplifies coloniality.

Without a protective policy, DWCC contracts would be issued in a manner that dovetailed with elite and state priorities, going to groups that best performed cleanliness and sustainability in its dominant aesthetic forms. This trend of privatizing waste commons and enclosing recyclables is proliferating across the world. In Uruguay, hermetically sealed containers prevent "classifiers" from accessing recyclables (O'Hare 2022). Dump modernization projects have threatened waste-picker livelihoods in Dakar (Fredericks 2022), Rio de Janeiro (Millar 2018), and Asuncion. In Indore, the municipality has contracted with a private company that collects waste in auto-tippers directly from households and transports it to a centralized facility where waste pickers are forced to pick through mixed waste and sell valuable materials back to the company at a fixed price (global_rec 2018). Similar dynamics have been observed in Ankara (Dinler 2016). In Delhi and Cairo, private companies displaced informal collectors from door-to-door collection (Calleja 2021; Fahmi and Sutton 2006). Bengaluru was headed in a similar antipoor direction with its attempts to formalize recycling.

Partial Knowledge, Scalar Mismatches, and the Burdens of Inclusion

Dry waste collection centers were to be instituted as public-private partnerships (PPP), in vogue in infrastructure provision in India and globally. The idea was that the municipality would provide the land and physical infrastructure for the center, while an industry partner (such as a corporate social responsibility [CSR] scheme) would provide financial investment and marketing, and an operator would take care of the day-to-day operations, including the work of cleaning, sorting, and transporting dry waste. Yet, very quickly, it became clear to middle-class advocates that corporate partners were hard to find. Reputational gains from corporate social responsibility were important to few companies, and it was challenging to obtain corporate sponsorship for all DWCCs.

Around the same time, groups representing the informal sector began raising objections: if recycling was to be institutionalized through DWCCs, it should not dispossess people already working with and living off waste. They accused the BBMP of favoring larger corporates over NGOs and informal workers (Nirupama V 2016). Parisara Tanda began to lobby TCGF and the BBMP to set aside some DWCCs to be run by waste pickers and scrap dealers. To counter the rhetoric that waste pickers were not good wage workers, Parisara Tanda argued that DWCCs should operate on an "entrepreneur model" that was economically efficient and quicker to establish. Such a model would leverage the autonomy and independent nature of waste pickers and scrap dealers, who would be willing to work in a low-margin business because they already did so. It invoked a kind of bootstrap narrative, arguing that if waste pickers were given an opportunity, they would surprise skeptics with their entrepreneurial hustle and business acumen.

But, for inclusion to take place, the state had to be convinced that waste pickers and NGOs could provide regular and reliable services. Initially, the eligibility criteria published by the BBMP required potential operators to have experience running a profitable business, making it nearly impossible for waste pickers or scrap dealers to qualify. Quickly, Parisara Tanda collaborated with some sympathetic TCGF members to draft memorandum of understanding (MOU) language with flexible provisions to encourage involvement by waste-picker collectives. It insisted that self-help groups and NGOs be exempted from this profitability clause. It also added a clause that operators should have at least three years of experience in waste management to prevent new entrants from capturing opportunities. Further, the MOU language stated that the BBMP should preferentially allocate DWCCs to waste pickers and scrap dealers.[24]

The MOU between the BBMP and waste pickers stipulates the roles and responsibilities of the two sets of actors (see table 4.1 based on an MOU signed in 2014). In this initial articulation, the eligibility criteria still indicated that operators were responsible for "raising resources to operate the DWCC," or, in other words, gaining access to operating capital. This stipulation functions as a barrier—waste pickers and scrap dealers have difficulty accessing credit at low interest rates from lenders. Further, the MOU clearly stipulates that the operator must meet operation and maintenance costs through the revenue generated from the sale of recyclables.

TCGF designed this low-cost DWCC infrastructure around "business principles," relying on a simple conceptual model: a clean city and maximum value extraction could be ensured via (1) consumer behavior change, (2) an efficient and well-managed DWCC, and (3) economies of scale. Embedded in this model were a series of ill-founded assumptions emerging from the individualistic and economistic thinking that dominates sustainability thought in this milieu (see also Pathak and Nichter 2021).

DWCC proponents assumed that simply changing consumer behavior would itself reduce the quantum of unmanaged waste in the city. This overemphasis on postconsumer sources of waste is a common malaise ailing plastic pollution campaigns (Liboiron and Lepawsky 2022, 40; Pathak 2020). The model assumed that if most households segregated waste, it

Table 4.1
2014 Memorandum of Understanding Terms for DWCC Operation

Municipality Role	DWCC Operator Role
Responsibility: Enforce segregation at source by residents and direct contractors to DWCC.	Responsibility: Create awareness programs about segregation at source.
Costs: Set up the DWCC by constructing a ventilated place with toilet and security. Facility should include security measures like gate and necessary beautification like sapling plantation and leveling of ground. Provide housing facility for security staff and printing signage and display board.	Costs: Pay a onetime security deposit of Rs 25,000. For waste pickers and SHG, Rs 5,000 payable over five installments.
Costs: Provision of electricity and water facility and payment of bills. Conduct repairs as needed.	Costs: Pay minimum of Rs 2 per kg of mixed dry waste to whomever brings dry waste to the centers. The center is not obligated to take any other waste, such as hazardous waste, toxic or post-production waste, or wet waste. Operator might choose to take wet waste and dispose according to the law.
Costs: Collect "reject" waste (sanitary, medical) for landfilling or incineration.	Revenue: Meet operation and maintenance costs through the revenue generated from the sale of recyclables.
Responsibility: Promote Extended Producer Responsibility takeback, buyback schemes. Assist operators in promotion of DWCC and awareness of segregation at source.	Responsibility: DWCC will display the purchase value of materials and will appoint labor for secondary segregation. It will ensure safety and health of workers. It will keep premises clean and take care of aesthetics, send monthly reports to the BBMP, and make documentation available to the public.

would eventually add up to a cleaner city, even though only 50 percent of municipal solid waste comes from residences.[25] Moreover, DWCCs were not appropriately scaled to keep up with the daily intensity and frequency of waste production in a city of Bengaluru's size and density. Most DWCCs were designed to manage only 1–2 tons/day (TPD) and the initial proposal was to build one DWCC per ward, adding up to a total daily processing capacity of 200–400 tons. With Bengaluru producing around 1,225 TPD of dry waste, there is a clear gap.[26] Operators needed to store or sort waste outside the premises to keep up with incoming volumes. This in turn led to these facilities being declared "eyesores," inviting the ire of neighboring residents (*The Hindu* 2016).

The bigger assumption was that infrastructural systems operate according to techno-rationalist rules: actors fall in line, efficiencies deliver value, and the state keeps its promises. ZWM leaders expected, or perhaps hoped, that the city's powerful garbage contractors would cooperate with this new arrangement, as it was in the public's interest. They anticipated that economies of scale enabled in the DWCC would generate adequate revenue: instead of selling small quantities, operators could amass larger quantities of recyclables, sell them to recyclers at better rates, and make the economics work. Finally, they expected the BBMP to keep its end of the bargain by constructing well-equipped DWCCs and making on-time payments for repairs and security.

These assumptions reveal three things: that when communal sustainability advocates became involved in designing urban infrastructures, they perceived waste flows as calculable, orderable, and predictable. Because all knowledge is situated and partial, middle-class zero-waste proponents lacked adequate understanding of the materialities of waste, the temporalities and social relations of waste work, and the subtleties of bureaucracy, resulting in a "scalar mismatch" (Liboiron and Lepawsky 2022, 39).[27] Importantly, they do not have to deal with the consequences of their utopian designs. Unlike waste pickers and other waste workers, zero-waste activists' knowledge of waste economies is secondhand. Their livelihoods do not depend on it.

Middle-class activists failed to consider on-the-ground relations among garbage contractors, sanitation workers, and waste pickers in their infrastructural designs. For garbage contractors, who until then collected mixed waste and transported it to landfills, the DWCC encumbered their business model and profit margins in three ways. First, garbage contractors rely on access to recyclable materials to avoid paying fair wages to the people they employ—they tell their employees to extract high-value materials and sell them to scrap dealers to earn a meager income. Second, the new system decreases the quantity of waste transported to the landfill, which would reduce the earnings of transport contractors whose fees were proportional to the weight of transported waste (*Deccan Herald* 2012). Third and finally, MOU language preferentially allocating DWCC opportunities to NGOs, trusts, waste pickers, and scrap dealers cut garbage contractors out of this economic opportunity.[28] Garbage contractors were unhappy with waste pickers and scrap dealers being included in a system that had been their domain. Some contractors mixed segregated waste at transfer points deliberately to undercut recycling attempts, while others refused to bring dry waste to DWCCs, cutting off the flow of inputs, which in turn compromised the viability of the business model.

Designers also underestimated costs and overestimated revenues. DWCC operators were expected to recoup operating costs entirely via the sale of waste. Maintaining steady inflow of mixed, saleable dry waste and a steady outflow of sorted, baled scrap was essential to sustain revenue streams. However, waste pickers did not have control over waste inflows. If contractors failed to deliver waste, or if sanitation workers extracted the highest-value materials, the DWCC would not have enough materials of value to recoup operating costs. Further, the price of recyclables fluctuates according to global trends in materials markets, meaning that recouping operating costs can take time. Operators must resort to taking loans to cover immediate costs or delay paying the sorters they employ. Moreover, because the plan did not account for hazardous waste, waste sorters were frequently exposed to toxicity in the DWCC when forced to handle broken CFL tube lights or sanitary napkins. These risks were significantly heightened during the COVID-19 pandemic. Waste pickers continued to operate DWCCs as an essential service during lockdowns and were exposed to health hazards from improperly discarded masks and gloves.[29]

A community radio show hosted by two scrap dealers on Radio Active Bangalore chronicled the lives and challenges of DWCC operators through ethnographic interviews. The two excerpts below show the risks and precarity these waste pickers face and communicate their commitment and persistence.[30]

> Maria and Krishna had to face consistent roadblocks after starting the DWCC three years ago. . . . Although they kept the DWCC functioning efficiently, they were always accused of inefficiency. . . . Through the center, over 30 tons of waste were being segregated, sorted and transported. Right when they had managed to convince the officials that they were indeed doing a good job, they received an evacuation notice. . . . Despite the residents being in support of Maria and Krishna, they had no choice but to shut the center down. With the center gone, the couple and the workers have had no source of income. Maria has been mulling about working with scrap dealers so that some income can come in.
>
> Aruvamudan and Cynthia Shiny, a couple who had been managing the dry waste collection of Ward number 170, 171 and 177 for the past 7 years have

been displaced by Namma Metro construction. . . . The DWCC had provided them with an automobile for collecting the waste as well as a center under the Jayadeva flyover for the purpose of waste segregation and storage. They collected about 800 kg to 1 metric ton of waste every day from about 700 to 800 houses. There is also a small team of women and men that works with them. It was tough going, but the couple was managing well despite the difficulties. . . . Now they had to vacate the center on account of the new Elevated Metro line that is being built. . . . They managed to find another space, but it is smaller and all the segregation needs to be done outside the center's main structure—a potentially grievous issue for the residents in the neighborhood.

Designers underestimated the ease with which the state lies to the poor and reneges on its promises. As the vignettes above show, the DWCCs given to waste pickers in most cases were no more than a shed, even though the MOU stipulated that the municipality was responsible for the provision of equipment such as weighing machines, baling machines, racks, tables, and chairs. Only 75 percent of the waste-picker-operated DWCCs have water facilities, and only 65 percent have toilets. The size of the centers varied greatly; some were so small that waste had to be stored outside, increasing the likelihood of theft. Many of the structures had issues with rainwater seepage and rat infestations. Some waste-picker-operated DWCCs have been gutted in fires, with foul play suspected (*The Hindu* 2021a). Further, many DWCCs are located far away from the wards they service because of NIMBYism from some middle-class groups who do not want waste processing sites in their vicinity, as well as land pressures. This means waste pickers must travel 15–20 kilometers daily for collection trips before coming back to the center to sort and bale. A time and motion study conducted by Resource Organization Collective found that waste pickers had to work up to ten hours a day to collect waste from larger wards in the city (Mathew 2022). These risks and burdens emerged directly from faulty assumptions about the material realities and social relations that were embedded in the design of the system. The DWCC thus became a place where waste pickers were simultaneously included but also faced the threat of superexploitation, a situation in which they must exploit themselves to keep the center running.

Materially Mediated Activism to Redesign Infrastructure

Despite these challenges, waste pickers refused to be passive executors of middle-class designs. Below, I identify two instances where waste pickers, through their embodied knowledge and materially mediated activism (White 2019), have leveraged their infrastructural inclusion to undercut neoliberal rollback and push for a more capital-intensive model that they designed themselves based on their expertise. The first of these was a policy decision made by the BBMP in 2018 that stipulated that waste-picker cooperatives or waste-picker-owned companies would have exclusive rights to operate DWCCs in Bengaluru.[31] Second, the BBMP conceded to paying a fee to waste pickers for door-to-door collection of dry waste, reversing the zero-subsidy model of infrastructure provision. These two policies, in addition to a 2022 decision to regularize sanitation workers—the outcome of a decade-long organizing effort and several strikes convened by unions representing *pourakarmikas*—might harken a "stealth remunicipalization" of solid waste management in the city.

Waste-picker organizations approached activism as a pedagogical task, engaging in a concerted program of popular education via Twitter, Facebook, community radio, street protests, and courtroom advocacy to show zero-waste activists, and the city writ large, that they had overestimated the recyclability of dry waste. The first set of advocacy actions was to quantify the flows of waste materials through DWCCs. For a period of several months in 2015–2016 and then again in 2021, waste pickers carefully tallied the exact amounts, material natures, and resale value of the materials that flowed into their centers. These accounts were made available on a website and shared with the BBMP. Twitter posts talked about how much pride waste pickers took in this act of quantification, dismissing accusations about lack of professionalism (see figure 4.3). This careful quantification exercise provided greater clarity on the actual material composition of Bengaluru's waste streams. Waste-picker records showed that a significant percentage of the waste entering their DWCCs every day, up to 80 percent, was low-value, nonrecyclable plastic waste. Much of this waste originates from the multilayer packaging that is found in fast-moving consumer goods. Readily recyclable waste such as PET bottles, glass, and paper made up a lower proportion of the flows entering the DWCC. To obtain one kilogram of recyclable material, a waste picker might have to bend a thousand times.

Parisara Tanda leveraged this careful documentation of dry waste to fight BBMP proposals to privatize DWCCs. In 2016, the BBMP was issuing tender requests for street-sweeping, household waste collection, and transportation. They included DWCC operation into the work package. This tender

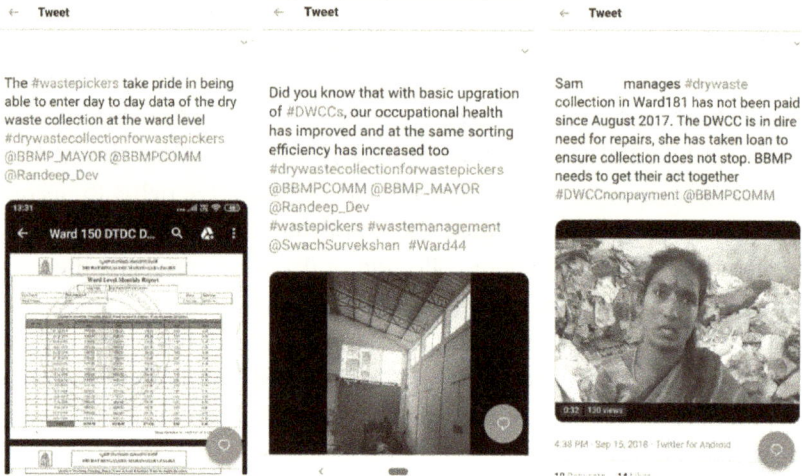

The #wastepickers take pride in being able to enter day to day data of the dry waste collection at the ward level #drywastecollectionforwastepickers @BBMP_MAYOR @BBMPCOMM @Randeep_Dev

Did you know that with basic upgration of #DWCCs, our occupational health has improved and at the same sorting efficiency has increased too #drywastecollectionforwastepickers @BBMPCOMM @BBMP_MAYOR @Randeep_Dev #wastepickers #wastemanagement @SwachSurvekshan #Ward44

Sam   manages #drywaste collection in Ward181 has not been paid since August 2017. The DWCC is in dire need for repairs, she has taken loan to ensure collection does not stop. BBMP needs to get their act together #DWCCnonpayment @BBMPCOMM

Figure 4.3
Screenshots of Parisara Tanda social media advocacy dated September 15, 2018, emphasizing the professionalism of waste pickers, as well as the ways the state has failed to keep to its end of the bargain. Reprinted with permission.

proposal would have immediately put waste pickers and other nonprofit DWCC operators out of a job, as the garbage contractor, a private corporation, would have taken over operations. Parisara Tanda and TCGF pushed back on this proposal, arguing that DWCCs were currently receiving mixed waste because the contractor had not equipped its employees to transport segregated waste. The garbage contractor had no incentive to encourage source segregation because the tender terms compensated the contractor for the deployment of assets and in proportion to the weight of different waste streams collected. Behaviorally, residents were much more likely to mix their waste if collection of different waste streams happened at the same time, TCGF argued. They recommended that the BBMP issue separate tenders for wet and dry waste. Parisara Tanda calculated extraction efficiency metrics using the data they collected in DWCCs and used this data to convince the BBMP that to increase resource recovery and recycling rates, DWCCs should be given over exclusively to waste pickers. In a complete about-face, the BBMP agreed to give contracts for collecting and processing dry waste exclusively to waste pickers and small scrap dealers, and even consented to paying them a service fee for picking up dry waste from households.

Waste-picker organizations also leveraged their knowledge of waste flows to work around middle-class obstinacy. The devaluation of waste-picker labor and faulty assumptions about the "wealth in waste" meant that bulk generators like apartment complexes sometimes refused to pay a fee for dry waste collection, even if mandated by government rules. Struggling to convince their clients that waste sorting is a labor-intensive task deserving fair compensation, waste pickers came up with a clever solution: "The middle class and rich people living in these gated communities refused to pay even 60 rupees/month service fee for dry waste collection and processing [for some context, a 300 ml can of Diet Coke cost 40 rupees in 2021]. Therefore, we came up with a pricing model to make them happy and not feel like they are paying too much. We moved from a fixed fee of 25 rupees to a variable fee. For any wet, organic waste, we charged them 2 rupees/Kilo. For dry, recyclable waste like paper, PET and HDPE bottles we paid them 2 rupees/Kilo. For rejects that cannot be recycled or composted, we charge them 3–4 rupees/ Kilo. Because most households produce so much reject waste, the cost actually comes out to about 80 rupees/month. So, this is better for waste pickers." Here again, waste pickers leverage their intimate knowledge of waste types and value to work around a system that was not designed to benefit them.

Finally, Parisara Tanda has consistently lobbied the BBMP to upgrade DWCCs with appropriate equipment. It has succeeded in having a small number of DWCCs enlarged to process 5 tons/day. Several DWCCs have also obtained collection vehicles from the BBMP through CSR schemes. In other cases, middle-class staff have cosigned loans with waste pickers to help them buy vehicles. Upgraded DWCCs have conveyor belts that reduce the need to bend constantly, balers that save space, and essential amenities such as toilets and a computer for record keeping. This equipment improves conditions of work but also supports the aspirations of waste pickers to learn new skills. From initially promising to operate DWCCs solely through their labor, therefore enabling the launch of the program sans any capital from the hoped-for corporate partners, to now slowly filling in the space in terms of adding more machinery and equipment and demanding more from the state, waste pickers are transforming these infrastructures utilizing a range of materially mediated tactics and strategies.

The Contradictions of Infrastructural Coproduction

Sustainability infrastructures are a productive location from which to "examine the constitution, maintenance, and reproduction of political and economic life" (Anand, Gupta, and Appel 2018, 4). While infrastructural undertakings, in concept and abstraction, have long promised development, progress, and security to governments around the world, the daily messy lives of infrastructures frequently undermine teleological narratives of technological progress, expanding equality, and economic growth. As opposed to seeing these twists and turns in the life of the DWCC as exceptional or as signs of failure, I suggest we read them as expressions of the contingent, dynamic, and emergent process of infrastructuring. Indeed, as anthropologist Nikhil Anand points out, waste infrastructures are particularly "flaky, falling apart forms that constantly call out for projects of management, maintenance and repair" (Anand 2015, as cited in Liboiron and Lepawsky 2022, 64).

In its ascendance and evolution, the DWCC gave material form to the shared priorities of elite environmentalists and informal waste pickers, strengthening their collaboration. Middle-class zero-waste proponents and waste pickers were united in their opposition to the spectacular infrastructure of waste to energy, but for different reasons. For the former, it was the threat of being made redundant, of unnamed and unseen toxicity, and of consenting to an infrastructural system that they could not monitor or control. For the latter, it was the threat of losing access to waste as the source of their livelihoods. Communal sustainability's emphasis on localism and decentralization and the entrepreneurial capacities of waste pickers met a municipality that was looking to devolve responsibility for keeping the city "clean and green."

Waste pickers realized the idealized imaginations of decentralized zero-waste infrastructure through their everyday work in the DWCC. In the process, they modify those visions as well—not through explicit political contention but through materially mediated redesign. It is through their labor that a nondescript shed or repurposed toilet becomes infrastructure that enables a productive and profitable city. Thanks to waste pickers, Bengaluru's ZWM could counter the perception that spectacular infrastructures are the only viable solutions to the garbage problem by demonstrating a decentralized, flexible, and quickly mobilized infrastructural configuration. In turn, becoming DWCC operators gives some waste pickers and scrap dealers improved incomes and more dignity as they come to be seen as waste professionals with valuable skills and knowledge to contribute (Ramani 2017). Several DWCCs are run by women, destabilizing the gendered hierarchy in informal waste economies where historically women, and Dalit women in particular, are relegated to the lowest rung of the waste-value chain, and have limited opportunities for upward social mobility (Sonal 2023).

Yet in the process waste pickers are exposed to new risks. There is less police violence but more financial risk. A majority of DWCCs are yet to be upgraded. The policy to give waste pickers and scrap dealers exclusive rights to operate DWCCs has never been fully implemented. The BBMP has failed to make regular payments to several DWCC operators, and unpaid dues now add up to millions of rupees. In 2022, the BBMP, under pressure to improve its standing in the Swachh Survekshan "garbage-free city" rankings, proposed appointing a single agency to collect all types of waste.[32] This move would mean waste-picker entrepreneurs can no longer collect dry waste directly from households, once again putting them at the mercy of private contractors. As in other cities like Pune, Bhopal, and Delhi, the state continually reneges on its promises to waste pickers, requiring constant displays of pleading, cajoling, and outrage to hold them to their end of the bargain. There are also internal hierarchies and clientelist relationships within DWCCs that merit more scrutiny—waste sorters employed by DWCC entrepreneurs do not always have access to fair wages, work contracts, or labor protections. They are more likely to be migrants or women from oppressed castes, and thus more vulnerable to gender and caste-based exploitation. The state's refusal to adequately compensate resource recovery work means that informal labor arrangements continue to subsidize formalized partnerships. Overall, the system continues to operate under what critical planning scholar Ananya Roy (2009b) identifies as the idiom of planning and urbanization in India: informality characterized by ambivalence, dysregulation, exemption, and opacity.

Tracing the DWCC reveals important insights about the opportunities and challenges of producing community-controlled infrastructures for municipal waste management. As Assa Doron and Robin Jeffrey assert, technological solutions for India's waste problems do not always need to come in the form of large incinerators or sanitary landfills (2018). Small technologies and improvised local solutions are important as well, especially when they are coproduced by diverse stakeholders and improve the conditions of work for labor. The DWCC is an interesting case study of the types of coproduced infrastructural solutions that can emerge when waste producers, civil society facilitators, waste workers, and administrators discover shared priorities and engage in collective action. However, its efficacy as a technology that "discards well" is undermined by scalar mismatch and sociocultural hierarchies that enable elites to pass off the burden and risk of operating infrastructure to less powerful groups. Its low-cost, provisional nature means that DWCCs can be dismantled just as quickly, leaving its informal operators adrift when policy priorities shift. Further, because the DWCC was designed without input from key constituencies such as the city's *pourakarmikas*, landfill-adjacent communities, or environmental justice groups, it failed to disrupt existing practices of disposability. The BBMP continues to rely on garbage contractors to dump unsegregated and low-value waste in Bengaluru's periphery.[33] Consequently, garbage contractors remain powerful actors in the city's waste politics, obstructing proposals from both environmentalists and waste pickers. Ultimately, this case demonstrates that infrastructural coproduction as a means of solving common problems in waste management is constrained by social hierarchies and dynamics of power that inhibit the full participation of minoritized stakeholders in collective action.

For the coproduction of waste management infrastructure to function not simply as a means to an end (and especially an end that primarily serves the needs of powerful groups), but rather to function as "a means of altering essential relationships and ongoing practices" (Mitlin and Bartlett 2018), it must remain a space of negotiation and subversion. Waste-picker organizations resisted some forms of formalization, privatization, and spectacular infrastructuralization, just as they sought to make other decentralized, labor-intensive forms more socially just by demanding more compensation for the work they do. They act as pedagogic agents, producing and sharing knowledge about waste and its management from their vantage point as waste handlers. Their knowledge improves both the efficacy of DIY infrastructures and their conditions of work within them. Yet, as municipal waste goes from being seen as a local environmental problem demanding better management to a global resource for solving multiple crises confronting consumer capitalism, the voices of both informal waste workers and local environmental groups become muted in policy deliberations. In chapter 5, I explore how urban struggles over the livelihood and environment in Bengaluru are connected to global flows and processes, implicating multinational brands, start-ups, and the global development sector.

5   The Win-Win Fallacy

I always knew going into business school that I wanted to solve real world problems plaguing developing economies. During one of my travels in India, the filth and squalor deeply disturbed me. However, I realized that something amazing was happening underneath. India was recovering and recycling nearly double that of any developing or developed economy in the world. But the benefits of such a system were not being felt. I wanted to solve all this and I wanted to build an organization that would fundamentally transform the way India saw recycling and plastics. This is how my journey from Silicon Valley to the back alleys of Hyderabad began.
—Banyan Nation investment video made by Impact Investment Exchange

"Informal, Illegal, and Largely Invisible"

Banyan Nation's investment video was made by Impact Investment Exchange (IIX), a Singapore-based organization that connects Asia-based social enterprises with impact investors. Banyan Nation, funded and supported by IIX, is one of many recent start-ups that have emerged in India's waste management landscape. The 2018 revamp of India's Solid Waste Management Rules to emphasize resource recovery and recycling, the launch of the Swachh Bharat Abhiyan, or the Clean India Mission, and a global push toward the circular economy has prompted young, educated, often foreign-returned Indians, like the ones featured in this video, to become waste social entrepreneurs. In keeping with a neoliberal ethos, these new enterprises deftly combine a narrative of doing good with a technologically driven approach to profit making.

If India already recycles more than other developed or developing economies, what problem is Banyan Nation solving? As one of the founders of Banyan Nation explains, "recycling activities in India are driven by market forces that are informal, illegal, and largely invisible. Millions of rag pickers today scavenge street corner bins and landfills . . . who then sell to the *kabadiwallahs,* who then sell to the back-end aggregators, who then finally sell to the recyclers. The goal of such an industry, really, is to recover the materials at the lowest possible cost, and at any cost." The images that accompany his words show industrious "rag pickers" going through municipal bins, working in the shadows of the night. Men quickly sort through and categorize materials by hand in dark and dingy rooms. He continues, "We started off by first tackling plastics. If you think about it, plastics is perhaps the most versatile invention of our time, but the single use nature of plastic means it has become an ecological and environmental poison. A lot of the plastics ends up in landfills. Whatever plastics is recycled is recycled in a very rudimentary manner. The back-end recyclers bite, bend, and burn plastic in order to identify the resin and grade of the plastic, forcing producers at the end of the value chain to depend on hundreds of such recyclers for consistent, quality recycling."

Invoking words such as filth and squalor and describing how workers "bite, bend, and burn," the video's narrative presents informal work as low-tech, polluting, and labor-intensive—the same narrative is used to advance formalization and privatization (Corwin 2020; Tucker and Anantharaman 2020). It argues that the dispersed, small-scale nature of informal recycling firms makes it hard for brands to source adequate amounts of recyclate, impeding an environmentally beneficial transition to a circular economy. The narrative also frames plastic as a necessary evil, and the cause of plastic pollution as poor management and inefficient recycling as opposed to overproduction. The video presents Banyan Nation's technological fix as it cues the dramatic music:

> The need of the hour right now is a formal recycling system, which ensures a superior quality and the ability to recycle the material that has entered the system more than once. To do that we have started off by building a very simple app in Hyderabad where we map over 1,500 stationary recyclers. This data gives us a bird's-eye view of the city of Hyderabad, data such as the amount of waste coming out of your house and data on the local efficiencies of the collection and transportation of waste. At Banyan, we have used mechanical and thermal testing techniques to produce a high quality recyclate that rivals virgin plastic. When the product reenters the waste value chain, its ability to be recycled increases by a factor of three. . . . Today we have established a proof of concept, we have integrated the supply chain, we are producing some world-class granules that

brands could use. We can scale not only across cities, but I also think we have the potential to scale across vertical.

Alongside this declamation, the viewer sees the Banyan Nation facility: two women wearing gloves, masks, and hairnets are sorting mixed plastics on a conveyor belt, while another feeds plastic pellets into a machine that then spits out a bright green resin. They work inside a large warehouse surrounded by orderly piles of discards. These workers are clearly not of the same social class as the founders, who, we are told, are engineers, MBAs, and former nonresident Indians (NRIs).[1] While the workers are silent, toiling figures in the background, the investment video lauds the commitment of Banyan Nation's founders. One suited investor suggests to the audience, "If you look at what [they] are doing, having given up what could have been otherwise extremely, I am sure, profitable careers, is to have built a technologically driven, technologically innovative organization." Another suited guy with an American accent chimes in: "Now you have a group of smart guys who are engineers or business guys, coming from Silicon Valley, getting advice from investors, trying to build a sustainable, scalable business." But the sacrifice is not in vain. Banyan Nation has gained a reputation after winning the Dell Circular Economy People's Choice Award at the World Economic Forum in Davos in 2018. L'Oréal, one of the world's largest cosmetics companies, and Tata Motors, India's biggest automaker, source recycled plastic from Banyan Nation.

Underemphasized in the video is an aspect critical to Banyan Nation's business model—its continued reliance on informal waste collectors. Banyan Nation argues that the need of the hour is a formalized recycling system, but it has integrated informal waste collectors into a mobile, digital, and Internet-of-things platform, through which it tracks waste flows in the city. It provides this tracked data to municipalities to help deliver data-driven waste management services. It employs waste sorters in its facility because manual sorting achieves a higher quality recyclate. In turn, it offers safe and sanitary working conditions, fair wages, retirement benefits, and health insurance for its workers. And, to informal waste pickers, it promises greater market access and better prices for reclaimed materials.

Start-ups like Banyan Nation claim to be vanguards of a transition to an inclusive circular economy. The circular economy is a global sustainability paradigm promising both environmental protection and continued economic growth through "closing loops." Supporters of this paradigm, from think tanks like the Ellen MacArthur Foundation to intergovernmental organizations like the World Bank to investment firms like Circulate Capital, are looking to partner with informal waste workers by offering what is described as a sustainable development triple "win-win-win": local and global environmental benefits (better waste management, less pollution, and reduced reliance on "virgin"[2] materials), social benefits (such as higher and more stable incomes for informal waste workers), and economic wins for corporations looking to secure access to material resources and adhere to increasingly stringent environmental regulations.[3] Is the circular economy finally a point of convergence for the interests of multinational corporations, municipal governments, middle-class performative environmentalists, and waste pickers demanding better livelihoods?

While infrastructural inclusion is often the hard-won victory of long years of struggle by waste-picker organizations, in this chapter, I warn that inclusion could be a double-edged sword. The win-win narrative of circular economies obscures a tension between profit accumulation and material improvements to the lives of informal workers. Because dominant sustainability discourse is replete with neoliberal common sense that promotes economic efficiency and market-based solutions to environmental and social problems, the outcomes for informal waste workers are differentiated and uneven. When initiatives privilege economic growth over social priorities, and waste pickers are patronized as "poor others" needing self-improvement, management, and supervision, *accumulation by inclusion* is the outcome. It subjects waste pickers to new regimes of discipline that can reproduce a dehumanizing caste-based division of labor. Thus, although transnational partnerships can help some entrepreneurial waste pickers enhance skills, improve livelihoods, and imagine social and career mobility for themselves and their children, these benefits are not available to most actors in the informal waste economy, who continue to be relegated to the role of silent, abject figures at the margins of global resource flows. In the coming years, as the circular economy policy agenda grows in prominence at both climate summits and in the negotiation of the global plastics treaty, Bengaluru's story issues a warning—economic inclusion in the absence of a clear commitment to social reform and reparation will reproduce racialized oppression.

The Waste-Less Frontier of Neoliberal Sustainability

Urban struggles over the environment and livelihood articulate with transnational sets of questions around how to restructure global capitalism in a time of overlapping crises.[4] Accelerating fears of resource shortages, the search for new productive frontiers of economic growth, combined with heightened awareness of how waste and plastic pollution contribute to climate change and biodiversity loss have coalesced around the circular economy. This term is applied to diverse grassroots practices, policy initiatives, industrial processes, and business models that try to move the economy beyond a linear "take-make-waste" model to one that "closes loops" by keeping materials in circulation as long as possible. Circular economy plans and programs have intensified the resourcification and commodification of waste at multiple scales, with more coordinated attempts to reclaim, repurpose, and recycle waste back into inputs for production (Kama 2015; O'Neill 2019; Savini 2019). Municipal solid waste is no longer simply a local externality to be managed, but a global resource and profit frontier to be mined.

The circular economy is an attractive proposition for corporations and national governments looking to "future-proof" capitalism by reconciling growth with environmental protection (Mah 2021). When China's National Sword initiative banned the import of plastic waste into what was until recently the world's plastics graveyard, consumers in the United States and Europe woke up to the stunning realization that their carefully sorted plastics were not being rebirthed locally but rather jettisoned off to faraway lands (Heiges and O'Neill 2022). The flow of "disposables-as-recyclables" (Liboiron and Lepawsky 2022, 71), which buttresses the status quo of throwaway cultures, was interrupted. As the alchemy of curbside recycling is demystified, concerned actors are paying more attention to where plastics end up. At the 2022 UNEA meeting in Nairobi, Kenya, member states of the United Nations committed to a legally binding treaty to address the plastic problem from production to disposal. Plastic waste producers are now under increasing pressure to demonstrate that they are making efforts to reduce the quantum of plastic pollution that enters our rivers and oceans. Curbside recycling schemes are no longer convincing performances of environmentalism. New props and actors must enter the stage.

Looking to assuage environmental anxieties and avoid binding regulations, corporations are turning to circular economy solutions. Some are promising to shift to reusable packaging materials, others to take back used packaging, and many more, like Coca-Cola and Pepsi, are committing to the easiest shift in their production processes: replacing "virgin" with recycled plastics. The move to use more recycled plastic is also an economic strategy to preempt legislation that could ban some plastics altogether (Liboiron 2013; Mah 2022). In their search for plastic and textile feedstock, big brands and circular economy proponents are turning to waste pickers, waste aggregators, and small-scale recyclers working in the informal economy in the Global South.[5]

It is in the informal settlements and back alleys of cities like Bengaluru that most global resource recovery work takes place. The informal recycling sector in Asia, Africa, and South America is estimated to be responsible for 58 percent of all the plastic waste collected and recycled globally (Lau et al. 2020). While the EU and UK might propose closing loops within their territories, their plastics are in reality often shipped off to poorer countries—in an expression of waste colonialism (Gregson et al. 2015; Liboiron 2021; Michaelson 2021).[6]

Informal recycling work has long been derided as polluting activity. But as geographer Melanie Samson reminds us, "It was reclaimers' knowledge and labor that first denaturalized waste, differentiated between waste and value, and expanded the waste-based commodity frontier" (2019, 68). Geographer Julia Corwin (2020) observes that environmental concerns about the hazards of e-waste have been used to secure corporate e-waste markets through devaluing informal labor. To date, most state and corporate-led recycling efforts have ignored or appropriated informal economies, sometimes instituting parallel resource recovery systems from scratch (just like Bengaluru's ZWM also tried to, until Parisara Tanda got involved). Many of these have failed because of successful protests by worker organizations. But they are also failing because it is hard to recover recyclables and make money from waste without waste pickers and sorters. Why?

Recovering recyclables from mixed waste is time-intensive, and it is not an easy industry in which to make a profit (Millington, Stokes, and Lawhon 2022). The informal economy can reclaim, divert, and recycle significantly larger quantities of material than the formal economy, particularly when it comes to reclaiming value in postconsumer waste (Gutberlet 2021). Informal economies have endured, despite persistent efforts to eliminate or formalize them, because of their "relative opacity, labor intensity and dependence on embodied knowledge (*metis*)" (Gidwani 2015), as well as through strategies of practical legitimation (Kornberg 2020). Informal actors are able to stay viable without much capital by charging marginal service fees for pickup services, relying on their embodied expertise and community resilience, as well as the capacity for self-exploitation enabled by socially reproductive relationships of care and coercion (Gill 2009). Manual sorting can deliver cost and material efficiencies that go beyond what centralized, mechanized sorting can provide.[7] In cities where formalization projects have been launched, informal work still subsidizes formalized recycling systems, resulting in what Patrick O'Hare describes as a state of quasi-formality (O'Hare 2020). This is why even technology-focused start-ups with ambitions of scaling up need waste pickers and waste sorters to build a viable business worthy of investment.

Accumulation by Inclusion

In the past few years, the governmental and corporate narrative on waste picking has shifted from looking to replace waste pickers, or "save" the poor from waste work, to instead looking to "include" or "integrate" waste pickers into new infrastructures and supply chains as a win-win proposition.[8] To understand this shift, I turn to Marxist economic geographers and what they tell us about how capitalism renews its conditions of possibility. Marxist theories of capital accumulation lay bare the inequities of the twinned processes of dispossession and inclusion that waste pickers are subject to in neoliberal cities.[9] Drawing on David Harvey's concept of accumulation by dispossession,[10] Vinay Gidwani argues that waste is a key frontier in capital's search for productive sites for its renewal (Gidwani 2013a). The (sometimes violent) enclosure of waste from an urban commons that sustained thousands to a tightly regulated resource controlled by private companies seeking profit and a municipality looking to cut costs has dispossessed waste pickers from their sources of livelihood (O'Hare 2022; Samson 2009; Schindler and Demaria 2019). The circular economy's focus on reclaiming the value in waste to future-proof capitalism puts these dynamics into overdrive, as impact investors flood the coffers of start-ups and set up parallel systems of resource recovery.

Responding to these waves of dispossession, organizations representing waste pickers in Bengaluru, Delhi, Bogota, Soweto, Belo-Horizonte, and many other places pushed for recognition of waste pickers' contributions to maintaining urban environments and cleaning the city.[11] This advocacy resulted in the integration of waste pickers into municipal schemes, alongside other modes of community participation.[12] Exploring how the state responds to infrastructural claims-making, geographer and discard studies scholar Rosalind Fredericks (2018) develops the concept of "governing through disposability" to describe how the neoliberal state takes advantage of people functioning as infrastructure, making ordinary residents responsible for functions that would have been in the ambit of public responsibility. Thus, while residents come to infrastructure to stake claims for citizenship and to renegotiate questions of dignity and justice, the state uses infrastructure to divert dissenting and activist capacities.

The state includes excluded groups, but in subordinated positions. The terms of integration are structured around individualizing, ahistorical, and depoliticized ideas about the poor and about poverty, alongside coloniality and casteism. As sociologist Manuel Rosaldo (2019) describes in studies of waste-picker integration in Bogota, waste pickers navigate and sometimes have to choose between dispossession and exploitation. Geographer Melanie Samson (2019) argues that many of Johannesburg's waste pickers reject or refuse integration on these terms.

Beyond a focus on urban infrastructure or urban livelihood struggles, I offer the term *accumulation by inclusion* to show how waste-picker inclusion efforts in Global South cities are becoming central to how consumer capitalism renews its conditions of possibility at a time of ecological crisis, environmental mobilization, and enduring poverty. Under capitalist social relations, accumulation by inclusion is usually thought of as occurring under the wage relation through the process of proletarianization.[13] But in a world where work outside the wage is as much the norm as the exception, and where social expenditure in maintaining a productive labor force has dwindled under neoliberal regimes, new forms of accumulation are proliferating. Glassman (2016) uses the term "accumulation by extra economic means" to name the enclosures and forms of accumulation enabled by the uncompensated but essential (gendered and racialized) work of social reproduction, internal to capitalism but outside the wage relation. Immanuel Wallerstein and other neo-Marxists like Nancy Fraser have pointed out that capitalism has always maintained a non-proletarian or semi-proletarian labor force to subsidize accumulation, just as in some cases workers have resisted their conversion into waged labor to maintain autonomy or avoid exploitative work conditions (Fraser 2021).

The social reproduction of the poor and of racialized others is not recognized as a resource until it becomes recognized as a resource for capital. Waste pickers have both resisted and been kept out of proletarianization. They have simultaneously subsidized capital accumulation through infrastructural labor while suffering abuse, violence, and humiliation. Social reproduction undergirds production and profit-making. Conscripting eco-activism and informal work into governmental programs through recognition and then devolving public responsibilities onto these recognized actors makes socially reproductive work available for capture—as we saw in chapter 4 with the case of DIY infrastructures. The state, start-ups, and multinational companies can obtain free or poorly compensated labor by creating inclusion schemes that play to the middle-class sense of self-importance as well as waste pickers' fear of dispossession.

Accumulation by inclusion is a strategy that keeps waste pickers performing infrastructural labor but with less autonomy than they had before. Policy and market interventions are designed to eliminate aspects of the informal economy that are seen as most problematic—the unprofessional and suboptimal nature of informal waste work through behavior change and self-disciplining. Such a deficit framing, when combined with caste as a force of humiliation, oppression, and control, renders waste pickers always subordinate in relationships. Thus the "extra-economic" means of accumulation operating here include the decriminalization of waste picking, the recognition of waste pickers as a useful governmental category invoked in policy documents, and the reconstruction of waste picking as visible, legible, and legitimate through surveillance and app-based tracking, as well as the reproduction of modes of discipline that are embedded in initiatives to include waste pickers into municipal infrastructures and global value chains. Through these extra-economic means, waste pickers become available as a flexible, cheap, globalized labor force for globalized resource recovery in the circular economy.

Following calls from relational poverty studies to "study up," instead of focusing my attention on waste pickers or the "poor others" targeted for reform and upgradation by global development actors and the state, I choose to critically examine ideologies and relationships of power and privilege that undergird efforts at waste picker inclusion. Global sustainability practice is emerging as a key arena wherein poverty and the poor are acted on. Circular economy initiatives linking global brands with informal workers function as "contact zones" in which poverty and the poor are problematized, and class difference is made and unmade. Examining the logics, ideologies, and imaginaries that animate inclusion efforts within the circular economy and other sustainability paradigms will help us understand why sustainability can reproduce injustice despite good intentions.

Most importantly, I bring attention to how the win-win narrative and the (a)politics of conviviality, which are endemic to most sustainable development interventions, obscure the enduring injustices experienced by informal waste pickers in inclusion arrangements. Indeed, that is the key point here. Waste pickers have long subsidized the conditions of accumulation by working with, alongside, and in place of formal waste management systems. The qualitatively distinctive aspect of this latest phase of integration is that it is presented as a win-win proposition that downplays the unavoidable trade-offs among economic growth, social equity, and environmental protection (see Liboiron and Lepawsky 2022, 138–144, on incommensurability). By recognizing and including waste pickers, exploitative and dehumanizing relations are repackaged in a manner that is more palatable to those working toward inclusive global circular economy transitions and sustainable cities.

I explore how accumulation by inclusion works through three modes: (1) through extended producer responsibility guidelines that recognize waste pickers but ignore the rest of the informal waste economy, retaining manual labor while replacing informal enterprises higher up the recycling chain; (2) through transnational initiatives connecting ethical consumers, brands, and waste pickers demanding legibility and transparency of waste-value chains and creating central roles for powerful legibility brokers; and (3) through app-based recycling programs linking waste generators and "worthy" waste pickers, which require the performance of self-discipline and self-improvement to access jobs and investment. Each of these examples reveal contradictions of inclusion in the context of the disciplining logics of profit-driven and market-oriented neoliberal sustainability.

Marketizing Waste, Governmentalizing Waste Pickers

Circular economy proponents promote the creation of new markets for waste materials, advocate technical retrofitting to improve efficiency, and propose new pricing models to internalize externalized costs—classic neo-liberal sustainability (Stahel 2016). This necessitates policy changes to waste management rules. In the 2010s, under pressure from environmental organizations, India's Ministry of Environment and Forests issued rules for plastic, e-waste, and municipal waste management that emphasized recycling and extended producer responsibility. These regulations recognize some sections of the informal economy as legitimate workers but deny recognition to others, which serves to retain manual work in last-mile collection and waste sorting while dispossessing informal enterprises higher up the recycling chain, thus making room for new private-sector entrants.

Rendering waste pickers governable through enumeration and registration is a precondition to integration into municipal systems or global value chains. In the 2016 Municipal Solid Waste Management Rules, Rule 11(1) (c) explicitly acknowledges the primary role played by waste pickers in reducing waste in India, while Rule 11(1) (m) mandates registration schemes for waste pickers. Rule 15 requires local bodies to formulate guidelines and create systems to facilitate the integration of waste pickers into municipal infrastructures.[14] The Swachh Survekshan urban cleanliness program that ranks major Indian cities on their visual cleanliness also awards points for informal sector integration efforts. Similar gains have been witnessed in South Africa, Brazil, and Argentina, with several cities issuing occupational identity cards and contracting with informal worker organizations for municipal waste management and recycling services. However, although waste pickers are now being recognized, India's policies, particularly the Swachh Bharat Mission, make no mention of scrap dealers, aggregators, and recyclers higher up in the waste-value chain. In India, elaborate and well-developed value chains exist for everything from hair and bones to toothpaste tubes.[15] But the state struggles to recoup tax revenues from informal transactions, preferring digitally facilitated commerce as well as GIS-based tracking and tracing, and therefore advocates the formalization of informal waste economies.

Marketization and formalization of informal waste recovery is promoted in the 2016 Plastic Waste Management Rules and the 2020 Guidelines on Extended Producer Responsibility. Extended producer responsibility (EPR), a concept originally introduced by environmental economists to hold brands and producers accountable for product end-of-life, is the bedrock of recycling and circular economy policymaking in Europe. EPR is supposed to incentivize producers to change their product design and manufacturing processes by forcing them to internalize the costs of recycling and disposal. In Europe, EPR implementation has taken the form of "take-back" schemes as well as legislation that imposes recycling fees on some products. The EPR guidelines in India clearly stipulate that the government will not fund or build public infrastructure for repair and recycling work. In keeping with neoliberal rollback of state responsibility for environmental protection, the government has restricted its role to simply facilitating market mechanisms.

Taking a cue from policy frameworks in Europe, the EPR Guidelines issued by the Ministry of Environment and Forestry in 2021 notified a class of "producer responsibility organizations" (PROs), intended to relieve industrial producers and commercial brands of their individual obligation to take back used products. Instead, the PRO fulfills these obligations on behalf of the companies it represents. Plastic producers and brand owners can contract with PROs to have them handle the collection and recycling of their own waste materials—for example, a beverage company paying a PRO to collect bottles postconsumption, which are then recycled in-house. Or brands can purchase "recycling credits" from PROs, wherein a particular percentage of the materials recovered and recycled by the PRO will now count for the brand having fulfilled its EPR obligation. This latter option is a plastics-offset credit system that is meant to work like carbon-offset credits—take no notice of the fact that carbon-offsets have been plagued by a range of technical and ethical issues (Hyams and Fawcett 2013).

Since the publication of these rules, a new generation of waste management companies are registering themselves as producer responsibility organizations with the central and state-level pollution control boards. PROs compete with informal value-chains by setting up parallel systems of collection and recycling. The national regulatory framework is more favorable to new entrants than to long-standing informal enterprises or collectives because the rules defining a PRO explicitly require the registrant to be a professional organization. Start-ups also often have an easier time accessing credit, as they are seen as worthy of investment by impact investors and venture capitalists. Importantly, national policy does not explicitly designate informal aggregators, traders, or recyclers (who are often denigrated as "middlemen"), leaving them open to dispossession. Consequently, waste pickers who previously sold their reclaimed materials to scrap dealers, aggregators, and recyclers in the informal sector must now interact with this new class of organizations led by elite, foreign-returned, or foreign individuals and funded by venture capital and impact investors. Market-based EPR mechanisms have thus empowered a new class of start-ups and social enterprises to replace or subsume informal operations.

Start-Ups and Social Entrepreneurs as Legibility Brokers

In August 2019, the UK fashion magazine *Marie Claire* carried a glossy feature, "The Body Shop Initiative That's Changing the Lives of Waste Pickers in Bengaluru."[16] A lead photo of five smiling women—whose lives we presume have been changed for the better—accompanied the celebratory headline. In subsequent months, more news articles appeared on this initiative, which sourced Fair Trade recycled plastic from informal waste pickers in Bengaluru.[17] A photo-essay in *Ethical Corporation* summarized the transformation of these waste pickers in one phrase: "From untouchables to plastic waste entrepreneurs."[18] In one photograph, you see a young Muslim woman sorting piles of plastics. In another, an old grandmother is laughing with Pixie Geldof, an English model and singer, hailed in the report as an "oceans' campaigner." Several photos featured waste pickers meeting Westerners, including representatives of Plastics for Change, "a technology platform that connects waste workers in developing regions with global companies."

Today, an increasingly crowded playing field of social enterprises, start-ups, and established companies claims to deploy digital technology and mobile apps to innovate waste recovery and recycling work in collaboration with informal waste pickers and scrap dealers. As waste becomes more widely recognized as a source of value and profit, new entrepreneurs as well as established companies are turning to this long-ignored sector (Doron and Jeffrey 2018). Companies like Kabadiwallah Connect, Recykal, Banyan Nation, and Plastics for Change function as *legibility brokers*. They mediate cross-scalar transnational relationships between informal waste pickers and global brands looking to demonstrate EPR compliance, and de-risk investments for venture capital firms. While these companies sometimes compete with informal economy "middlemen" such as large scrap dealers, aggregators, and recyclers, they simultaneously integrate small scrap dealers, informal waste pickers, and itinerant buyers into their platforms for last-mile waste collection, manual sorting, and other labor-intensive, low-margin activities that resist automation. Although modes of integration vary, included waste pickers are subjected to new governing logics and relationships as they navigate a new set of actors.

Legibility brokers espouse a dominant logic in the sustainable development literature in which informal recycling is labor-intensive and therefore has much potential for increased "productivity" through technological upgrading and innovation (Schroeder, Anggraeni, and Weber 2019). However, while many start-ups claim to be upgrading technology, I find their primary interventions are not always in the form of better machinery (though there are some examples of this), but rather in the form of information as technology. Start-ups seek to ease transaction costs by connecting waste producers, waste pickers, plastic granule producers, and brands through mobile apps or digital platforms. While some provide waste pickers with space to store and segregate plastics in exchange for a franchisee fee (basically landlords collecting rent), and others have innovated washing technologies that deliver a higher-quality recyclate, the main value proposition start-ups offer to the global circular economy is assurance of compliance, traceability, and transparency.

Circular economy proponents argue that upstream and downstream visibility is key to both encouraging and demonstrating circularity (*Wired* 2022). If consumers and companies know what materials are in their products and where their waste is going more responsible practices will emerge, goes the facile logic. Brands looking to demonstrate EPR compliance to regulators also demand transparency and trackability from their suppliers. To help brands demonstrate headway on their CSR goals, some start-ups also offer cause-marketing support and customized impact measurements of CO_2 savings, livelihood creation, and progress toward sustainable development goals. Others have obtained certifications, such as fair-trade certification from the World Fair Trade Organization. Companies like Plastics for Change also sell plastic offsets, promising to interrupt the flow of waste into oceans on behalf of businesses and consumers that buy their offsets.[19] Producing these metrics and enacting these market-based arrangements requires a lot of monitoring.

To make informal waste recovery legible and legitimate to global brands, start-ups often have in-house sourcing teams that recruit and monitor informal waste collectors. Monitoring is achieved through GPS tracking and digital banking, as well as on-the-ground support using behavior-change incentive models wherein waste pickers get better rates for better segregation practices. Start-ups ask their waste picker partners to supply them with a steady stream of segregated, clean, high-quality plastics like PET and HDPE (they usually do not want the low-value materials, which waste pickers must dispose through other means), collected under "dignified work conditions." This means plastics should be collected directly from source (as opposed to picked from a dump), a location should be designated for the sorting of plastic (so, not in homes or on the street), and there should not be any children in the vicinity of the work site.

Community care work and household reproductive work subsidizes this new frontier of corporate greening. Community-advocacy by middle-class environmentalists and waste pickers improved source segregation, making it possible for start-ups and waste pickers to source plastics directly from households. Start-ups often rely on donor-funded programs to provide the social services that sustain dignified labor conditions. For instance, a Community Library network operated by Parisara Tanda cares for the children of waste pickers while their parents sort waste at aggregation centers. Plastics for Change has an eponymous foundation that provides emergency support and community development services to some waste-picking communities.[20]

Self-Improvement in Place of System Change

The global circular economy agenda promotes market-based mechanisms to reduce pollution and increase resource recovery. In this model, polluters are asked to pay to upgrade recycling infrastructures—either directly as impact investors or indirectly through the purchase of EPR credits. To gain access to these flows of money that they need to improve their livelihoods, particularly in the absence of support or investment from the state, informal waste pickers and recyclers are forced to compete or collaborate with start-ups and pursue partnerships with global brands like Coca-Cola. To present waste pickers as worthy of inclusion into regulatory frameworks and global partnerships that favor formal enterprises, organizations working with waste pickers focus on professionalizing waste picking. If waste pickers and scrap dealers demonstrate business and technology skills, they will be taken more seriously by policymakers, brands, and investors, and be better placed to take advantage of the new economic opportunities of the circular economy, goes the logic.

Calls for professionalization emphasize the vulnerability of waste pickers alongside perceived inefficiencies in the system. As the director of a charity working with child waste pickers explained in a 2012 interview,

> The context of India where we have got a very strong robust presence of informal sector both in terms of collection and in terms of, you know, the skillset. But at the same time, we also feel that whatever is happening in the informal sector right now, we feel that it's not adequate, it's not visible, it's also not functioning at its optimal level, ensuring that the people who are involved are protected from the vagaries of this waste management. What I mean is that any fluctuation in the market price of the recyclables can affect the lives of these people. Any policy changes, any new technology comes, for all these things these people are vulnerable. So therefore having a kind of structure or any kind of mechanism within the waste management, which is formally recognized and financed for as well as given the legitimacy of operating, we felt it was very important.

But while most actors identify systemic issues around lack of financing and regulations, waste pickers themselves remain the primary targets of intervention for professionalization. The NGOization of the waste-picker social struggle,[21] as well as an increasingly hostile political climate for Dalits and Muslims in India, means that more radical critiques are deemphasized in place of a self-improvement ethos.[22] Partially this is also due to the donor-funded context of work with urban poor, which tends to favor projects that make populations, as opposed to systems or social structures, the targets of intervention. Anthropologist Tania Li (2007), who studies social development schemes, names this as a quintessentially neoliberal approach that focuses on encouraging people to take responsibility for their own improvement by gaining skills and making the right choices to engage in competitive marketplaces. National regulations and municipal bylaws also favor private companies over cooperatives as service providers. Parisara Tanda founders cite this as the reason they started a social enterprise, which is authorized by the BBMP to provide waste management services in Bengaluru. This company enacts the entrepreneur logic: individual waste picker–entrepreneurs can become franchisees of this social enterprise and then contract directly with apartment complexes, gated communities, businesses, and commercial operations to provide waste management services. Waste pickers who become franchisees (usually men) in turn hire other waste pickers as sorters (often women or migrants) and drivers (mostly men) to work for them. These labor arrangements are often ad hoc and informal.

To make themselves includable in municipal infrastructures and global value chains, waste pickers are advised to change their appearance and behavior and to build their competencies (Luthra and Monteith 2021).[23] A promotional video for the now-defunct I Got Garbage platform run by Mindtree, a large multinational technology company, captures this invocation quite explicitly. Through the I Got Garbage web platform, bulk generators of dry waste such as apartment complexes can request waste pickup from informal waste pickers. The waste generator can either sell the waste to the waste picker at a fixed rate negotiated by the Mindtree platform or choose to donate it. The platform also required digital payments and invoicing, which brings previously untaxed cash flows into the realm of regulatory and tax oversight.[24]

The video starts by showing us a disheveled and unshaven man picking in a landfill and living on the streets, feeding scraps to his child and a stray dog. He happens to go on a computer and clicks on the IGG website. Like magic, he is transformed into a smiling, clean-cut worker. He now has a truck to collect segregated waste door-to-door. He goes to the bank to get his earnings and can buy his daughter (who is now seen carrying a school bag) ice cream from a vendor. He is now accepted, even well regarded, in society. But what the video, and more broadly this simplistic narrative of change, leaves out is that the truck is likely obtained through a bank loan at a high interest rate.[25] And what if payments from the government or the company he works with are delayed—as they often are? Also, where does this newly included waste picker live now? At a time when informal settlements have been targeted for removal in Bengaluru, has the waste picker escaped the slum or has their shanty been wiped away by floods? Parisara Tanda staff recount several instances in 2019 when waste-picker communities living in slums were targeted for expulsion. They were accused of burning garbage by the very middle-class actors who celebrated the inclusion of "worthy" waste pickers.[26]

Even integrated waste pickers who have demonstrated professionalism face moments of discipline and humiliation. Geographers Aman Luthra and Will Monteith write about how waste pickers and street vendors in Delhi use uniforms and codes of conduct, alongside identity cards, to legitimize themselves in the eyes of the state and capital (Luthra and Monteith 2021). ID cards and uniforms offer a degree of dignity and protection from abuse, but this legitimacy is fragile and can be retracted at any point, as this vignette shared with me by a Parisara Tanda staff member illustrates:

> While skimpily clad cheerleaders gyrate in support of the city's football club, waste pickers like Sagaya and Hari work the event, going through the crowd to pick up and separate plastic, paper, and glass. They work quickly and discreetly. The event organizers appreciate their speed, efficiency, and unobtrusiveness. However, they were less pleased when Sagaya hitched up her *sari* to keep it clean. While cheerleaders, many of them white women, flashed their midriffs and thighs, the sight of a waste picker's calves was deemed inappropriate for the venue. The organizers complained to Sagaya's employer. The hitching up of her *sari* meant that she was no longer invisible or in the background. No longer a uniformed worker in her place, carrying out her jobs.

Undercutting Solidarity and Collective Power

Neoliberal inclusion models support individual success for some entrepreneurial waste pickers at the expense of collective benefit for broader populations. These arrangements are extended to receptive individuals with the right dispositions for business. Indeed, some waste pickers have made it, achieving higher incomes, better work conditions, and more security through participation in these new value chains. One example is Deepak, who has started his own company that provides waste management services to several households and commercial enterprises in southeast Bengaluru. Deepak was a child waste picker who has emerged as a leading voice for waste-picker rights in the city. He also runs a dry waste collection center, is featured on a global brand's website, and speaks on national waste management panels. Similarly, Imran, another entrepreneur, conducts zero-waste trainings for middle-class communities, holding his own in elite-dominated spaces. Periamma has operated a DWCC for more than eight years and has built her own home. But as Deepak reminds us frequently in his public speaking, there are thousands more waste pickers like him, whose lives are yet to improve. All while private corporate actors enter the waste recovery and circular economy space, outmuscling waste pickers with swanky investment videos, English-language skills, and their economic and cultural power.

Entrepreneurial opportunities and acumen might allow a select number of waste pickers to escape stigma, but it leaves the stigma, humiliation, and precarity associated with both poverty and caste generally intact. Akin to what Patrick O'Hare (2022) observes in Uruguay, a cleavage is emerging between organized, included waste pickers and waste pickers who still operate outside these inclusion initiatives. In Bengaluru, only about 1,500–2,000 of the city's 15,000-odd waste pickers have participated in livelihood inclusion programs such as working at DWCCs or providing waste management services to apartment complexes.[27] For waste pickers operating outside these inclusion initiatives, the enclosure of waste materials has reduced their access to recyclables on the street. Although identity cards offer a degree of protection and dignity, they remain at risk of facing the quotidian harassment and casteist stigma that was once the fate of all waste pickers in Bengaluru.

These new integration arrangements can also undercut preexisting relationships of mutual tolerance, solidarity, and dependence that buffered informal workers from the vagaries of the market and elite opinions. A community journalist and waste researcher who has studied inclusion arrangements across the country explained in a 2021 interview that inclusion arrangements can undercut waste-picker agency:

> In waste-picker colonies in Bengaluru where dry waste is aggregated, the waste pickers have some agency on whom to sell the waste to and can sell it to the secondary aggregator or recycler who offers the best price. In these new relationships, especially when the Urban Local Body or waste contractors are involved, waste pickers are forced to sell to them at a much lower rate than they could get on the open market. Further, in some of these new companies where waste pickers are getting jobs, they do not have freedom of association, collective bargaining, or fair remuneration. There is limited oversight, and therefore a lot of exploitation behind the scenes. Moreover, a lot of these new companies or even the state are coming in to get rid of the middleman, the scrap dealers, and aggregators. However, in many cases, these informal recyclers and aggregators are also the actors who provide a buffer to waste pickers during times when the price of plastic comes down, by either providing credit or space to store waste. They serve as the safety net for the waste pickers. Finally, there is a limitation of how many waste pickers can even get jobs in this new, parallel waste economy.

The integration of waste pickers into corporate-controlled recycled value chains can also instrumentalize their labor and knowledge.[28] Urban recycling work has been resistant to capture because of the unique expertise and embodied knowledge (*metis*) held in these networks and communities. But the emphasis on transparency seeks to leverage and control where and how *metis* for value extraction is deployed, without adequately compensating it. The tacit knowledge of waste pickers regarding where to collect and how to sort is available for capture through GPS tracking devices that monitor their movements, and by forcing them to commit to supervision. But the knowledge of technology entrepreneurs innovating in the circular economy is protected as proprietary via patents. And indeed, that is the other danger here—that, as Melanie Samson describes, the commodity frontier first created by waste pickers is now ripe for capture by more powerful and well-resourced private actors in act of epistemic injustice. Dispossession might be the outcome of inclusion after all. Thus, when the Ellen MacArthur Foundation asserts that states and businesses should be "tapping activities of the informal economy" (2016, 28), to hasten a transition to a circular economy, it remains to be seen whether they will be tapped dry.

Accumulation by inclusion is a strategy by which capital devolves infrastructural labor onto poorly compensated laboring bodies in exchange for removing the threat of more severe dispossession. By rebranding self-exploitation as entrepreneurism, this strategy effectively silences dissenting voices questioning this form of integration. Integration is not a new thing. Private contractors or the state in several parts of the country subcontract with waste pickers who, in exchange for access to waste, do the demanding work of door-to-door collection and waste sorting. Some of these arrangements are formally negotiated (such as in Pune or Delhi), while in other cities the arrangements are unofficial and sometimes more exploitative. What is qualitatively distinctive about this latest phase of integration is that it is presented as a win-win proposition under a broader global sustainability paradigm. There are enough success stories to write about in glossy magazine pages. Yet, these images distract from the enduring suffering and exploitation of many waste pickers. This model does not need everybody to succeed.

Reducing Costs Will Reproduce Caste

To make waste recovery more efficient and less dangerous, waste picker organizations and entrepreneurs seek space, equipment, social security, and guaranteed payments, both from the state and from private corporations. But corporate and state collaboration with the informal economy is largely motivated by a desire to secure the conditions of growth and profit. They consequently do the bare minimum to capture value from informal work— as also seen in chapter 4.

Cheapening is an economic and cultural strategy employed to exploit waste pickers.[29] The economic strategy involves undercutting the collective power of waste pickers through subcontracting and casualized labor arrangements, which forces them to take on risk, keeps them out of the wage relation, and denies them labor protections. The sociocultural strategy is focused on maintaining caste as an unspoken but ever-present force of humiliation and stigma to silence, coerce, and discipline waste pickers. Caste twins the essential and disposable in Indian cities, reproducing certain spaces and categories of people as sinks, buffers, and shock absorbers. Structures of humiliation keep waste pickers, as mostly Dalit women and men, migrants, and religious minorities, locked in positions of abjection (Reddy 2015). This unequal power relation in turn makes it harder for waste pickers to demand higher wages, more decision-making power, or more investment in their own reproduction through education, training, and housing support.

Because waste-picker organizing in Bengaluru is filtered through the elite and dominant-caste zero-waste movement and constrained by a market-based regulatory regime, an emphasis on efficiency, managerialism, and self-improvement triumphs over attempts at social reform to abolish caste. The participatory decision-making forums that many waste-picker advocates globally have argued should be at the core of all efforts at integrating informal sector actors have not been set up, and no plans to do so are appearing in the short term (Carré, Horn, and Bonner 2018; Dias 2016). Decisions are still made by "expert committees" and technical bodies in lieu of democratic fora. A longtime environmental justice activist in Bengaluru, captures this sentiment trenchantly:

> Caste has replayed everywhere, even now if you see the situation, the waste pickers are obviously the poorest of the poor. And even if you integrate them really, there's really no social reform, is there? . . . That's because if you get the people who run the organization, they're largely a product of middle-class education and upbringing. So unless you train yourself and school yourself, you are stuck within that. You are still within the efficiency management paradigm.

A zero-subsidy, profit-oriented model cannot help us "discard well." It will only make waste management seem more modern while retaining its most backward parts. To be clear, the intentions in most of these partnerships are often benign: to improve conditions of work, reduce or eliminate toxic exposure, or increase efficiency of resource-recovery using technology.

These interventions are also necessary. It is important to emphasize that some waste management tasks, especially those involving wet and mixed waste, can and should be mechanized, and technology plays a key role in that. Casteist municipal infrastructures condemn manual scavengers, conservancy workers, and waste pickers to bodily toxicity and social humiliation to keep spaces clean—demanding human contact with refuse reduces costs (Sreenath 2019). Mechanization and deployment of technology are necessary to break this relation, in addition to sociocultural change. But for now, technology development and deployment is done with the goal of sustaining growth and extracting profit rather than ending casteist oppression in waste systems. Indeed, manual labor is retained when it supports profit-making.

Casteist ideas of intellectual and moral worth also mean that, once waste pickers are included in the waste management system, they must go to great lengths to demonstrate that they are hardworking, efficient, and reliable. On the other hand, foreign-returned, English-speaking, and dominant-caste techies are already seen as saviors making great sacrifices by descending into this dirty business. This dynamic has caste written all over it. But casteism is constantly denied. A BBC study conducted in 2021 revealed "55% of the study respondents felt that informal waste pickers were dirty in appearance. A substantial 56% even believed that they should not be allowed in building complexes and societies. The strong negative perception was attributed to the physical appearance of waste pickers, impacting their interaction and acceptance in society" (Kappan 2021). Casteism is not acknowledged, and the waste picker is blamed for their poor appearance.

Accumulation by inclusion is a strategy pursued by state and corporate capital and enshrined in national and global policies of the circular economy. It emerges out of the realization that waste pickers are the cheapest available option for waste collection and sorting, essential tasks for materials recovery that are hard to automate. Treating waste pickers as a globalized resource for the circular economy enables municipal governments to cut costs and national governments to meet climate commitments and secure resource access. It gives new entrepreneurs a cheap labor force and multinational brands a CSR talking point. Along the way, it also salves the green guilt of consumers discarding plastics who also want to do their bit. But the bodies of people who are locally stigmatized, othered, and humiliated are the fulcrum of this emerging transnational flow of discourse, money, and material.

Moving from Economic Inclusion to Social Reform

I am at the 4th United Nations Environment Assembly (UNEA) in Nairobi, held in March 2019.[30] These assemblies were created after the Rio 2012 Earth Summit to ensure regular deliberation and progress on global environmental challenges. I am attending as an invited speaker at a Future Earth event on "Enablers of Sustainable Lifestyles." I have a credential that gives me access to all the events. One of the focal themes for the 2019 event is marine plastic pollution. Consequently, there are several sessions dedicated to waste management, plastics governance, and the circular economy. I decide to attend a session on finding alternatives to plastics. The panelists include a representative from Coca-Cola, one of the world's largest producers of marine plastics pollution, and another from BASF, a petrochemical company that manufactures plastics. Also in the mix, a government representative from the Ministry of Environment in Kenya, which had just enacted a pioneering single-use plastics ban in the country. Kenya, along with India and the EU, were pushing for UNEA to adopt a commitment to phase-out most problematic single-use plastic products by 2025.

The government representative, who speaks first, says the plastic ban was the only way to get the attention of industry and get them to cooperate. His comments firmly place the onus for further action upon the industry representatives, who speak next. Coca-Cola and BASF predictably point to the charters they have signed and their ad campaigns as evidence of their commitment to ending plastic pollution. They promise to use more recycled plastics, develop biodegradable alternatives, and increase efficiency. They simultaneously deflect attention and responsibility onto the consumer and onto public infrastructure. "Consumers need to do more to recycle, and local governments need to set up better collection systems, the informal sector needs to be formalized" says the BASF representative, deftly taking attention away from their role as a major plastic producer. Coca-Cola absolves itself of responsibility for the plastic pollution crisis by instead blaming "the current plastic collection and reuse system [which] is broken." (Coca-Cola Company n.d.)

In the Q&A session, an 8-year-old is introduced to the audience as a young environmental champion. This young Kenyan directs a question to the Coca-Cola and BASF representatives. "My parents say that when they were young, they used glass. Now there is plastic everywhere. Why can't we go back to a system where we use glass bottles that are more easy to recycle? Do we really need plastics as you seem to say we do?" The panelists look somewhat stunned by the clarity of the question. Finally, the BASF representative speaks up in a patronizing tone "Plastics are better than glass, more durable and easy to transport. Glass can break which is dangerous. Right now we don't have biodegradable plastic that can hold water, so until then we have to use plastics, if not, you will not be able to get the Coca-Cola, and that would be bad, right?" Coca-Cola is rendered an essential need in her response.

A few days later, the text for the first global commitment on "Curbing Single-Use Plastics" is released. Lobbying by the US, Saudi Arabia and other countries with ties to fossil fuel and big petrochemical companies weakened the resolution (Bhalla and Ndiso 2019). They watered down the language from "phasing-out" to "significantly-reducing" single-use plastics. Brian Doherty, a member of the U.S. delegation at the UNEA, convinced other delegates that rather than focusing on phasing out single use plastics, the emphasis should be on fixing waste management in countries that are major sources of marine plastic pollution. "We support reducing the environmental impacts from the discharge of plastics, but we further note that the majority of marine plastic discharges comes from only six countries in Asia where improved waste management could radically decrease these discharges" he said.[31] Once again, the focus is successfully deflected from corporations that push over-consumption to brown bodies laboring over waste.

Plastics are on track to contribute more greenhouse gas emissions than coal plants by 2030 (Valette 2021). Even though plastics are implicated in a range of environmental problems from toxic pollution to marine biodiversity loss, the global plastics industry has managed to avoid stringent regulation for decades now. From the United States to Bengaluru, companies pose plastics as a necessary evil that can be made less evil by changing consumer behavior, instituting recycling programs, and, now, through the market-based solutions of the circular economy.[32] This dissimulation is becoming more pervasive because plastics is the fossil fuel industry's ticket out of redundancy.[33] Petrochemical companies are looking to circular economy "solutions" to both contain environmental crises and create new frontiers of accumulation (Mah 2021).

At the Fourth UNEA, I witnessed firsthand one strategy the plastics and petrochemical industry uses to evade regulation—deflecting blame onto consumers and informal workers. This is an act of dissimulation, backed by PR campaigns and advertising agencies. It is meant to make consumers blame themselves for a problem they have had little part in creating. By pointing the finger at "only six countries in Asia," blame for this global environmental problem is placed on informal workers and waste management systems in the Global South. Today, these very companies invoke the circular economy as a "win-win" solution, downplaying the trade-offs between economic growth, social equity, and environmental protection that are inevitable in any intervention. If the circular economy policy agenda is to be recovered from the techno-managerial shackles placed on it by powerful corporations and their allies, questions of environmental justice must be centered, as must the political voices of informal workers.

Three years later, at the Fifth UNEA in 2022, member nations have finally agreed to create an international, legally binding treaty on plastic pollution by 2025. For this new plastics treaty to be effective and cohere with the Sustainable Development Goals, waste pickers and other informal waste actors insist that they must have a seat at the negotiating table and a role to play in tackling plastic pollution (GRID_Arendal 2022). Waste pickers were present at the summit, and thanks to their activism, the summit's resolution included language "recognizing the significant contribution made by workers under informal and cooperative settings to collecting, sorting and recycling plastics in many countries." The first battle for recognition has been won, but now the longer struggle for just remuneration, full participation in national and global policymaking, and a just transition begins.[34]

On the other hand, a top-down inclusion agenda tends to see informal workers as an untapped resource, an opportunity to meet recycling goals at least cost, simply because this is what corporations always do—cut costs. These agendas are not readily reconcilable. When companies claim to balance economics, environment, and equity, it is usually the last "e" that loses out. To prevent inclusion from becoming repackaged oppression, treaty negotiations must begin with the acknowledgment that current systems of disposal and recycling represent waste colonialism and environmental racism/casteism.[35] This is expressed not just through the forceful appropriation of land as environmental sinks, but also through the systematic subjugation of bodies through racialized regimes of labor for recycling. It must also recognize that relying on market mechanisms such as plastic offsets or EPR mechanisms will reproduce this racialized division of recycling labor, especially if workers have limited ways of holding companies accountable for unfair inclusion terms. There is no level-playing field here. Even a well-meaning inclusion agenda can reproduce oppression if it fails to consider the deeply cultural and embedded social hierarchies that sustain current systems of discard. In the concluding chapter of this book, I articulate an alternative agenda of "reparation by inclusion," which requires the circular economy and other sustainability paradigms to go beyond a narrow economic rationality to reemphasize mutual vulnerability through transformative coalitions.

Conclusion: Beyond Inclusion, toward Reparation

In a 2020 advocacy video produced by Parisara Tanda, Deepak, a former child waste picker who now operates a dry waste collection center (DWCC) in Bengaluru, articulates how market forces affect waste-picker livelihoods. The video was released to offer waste-picker perspectives on the extended producer responsibility (EPR) guidelines issued by the Indian government earlier in the year.

> They [waste pickers] work from morning to evening until their body aches. The amount of money they make compared to the effort they put in is very little. This happens because the market behaves like that. . . . When there was demonetization,[1] there was no money, so no trade could happen. When there is price fluctuation of crude oil, it will affect the price of the recycling market. We have gone through a lot of ups and downs, a lot of struggles, there are no predictable rates we get for the material. . . . When there is a demand for recycling plastic, they give one rupee more. And when there is a surplus recyclable ore, give us two rupees less. However, the hard work we put in is the same, there is no difference. EPR legislation requires companies to take back their waste, . . . but I ask those companies, please think before you implement these strategies. If you appoint another company, they might not understand the grassroots. Please work with us directly. Do not exclude waste pickers. The important thing in EPR is the collection of data. That data, we can ourselves give it to you. We have data on what materials we have collected, the rates at which we have sold it. We have learned to use a smartphone, we have developed an app. We are using many technologies. We are ready to learn and use any technology required.

Looking for Transformative Potential in the Cracks

Can sustainability and environmental action be emancipated from neoliberal capitalism's eviscerating impulses? Deepak still thinks so. In his comments, he outlines some tangible policy actions to improve the lives of waste pickers. He calls for fixed rates and fair prices that protect waste pickers from the vagaries of global markets. He asks corporate brands to work with them directly and avoid the seduction of the latest tech-speaking start-up to come into town. Deepak's comments go beyond policy prescriptions. Fundamentally, he is asking the state, consumers, and corporations to overcome their prejudices. Implicit in Deepak's demand is the recognition that the system as it currently stands is unfair. That it devalues waste pickers' labor while taking it for granted.

For waste pickers like Deepak who were already doing the infra-labor of maintaining urban environments, inclusion into decentralized sustainability infrastructure is both a burden and an opportunity. Deepak is now required to collect inorganic dry waste from several neighborhoods in Bengaluru as part of his contract with the Bengaluru Municipality (BBMP). Much of this waste does not have a preexisting market, and the resale price is always fluctuating. Nevertheless, he is now inside the system, and from his vantage point and by leveraging his knowledge, he can attempt to participate in policy discussions. There is an ongoing conflict here, simmering just under the pile of discards in the background of Deepak's video. On the one hand, there is a top-down governmentalizing goal to make waste pickers perform recycling work for as little compensation as possible. On the other hand, waste pickers try to both resist and modify these schemes to advance their goals of higher, stable incomes and social mobility. And surrounding this conflict is the broader question of how to deal with Bengaluru's discards in an ecologically sensible manner.

The top-down program that repackages the exploitation of waste pickers is what I call a neoliberal sustainability agenda. As best exemplified by the corporate-led circular economy, it is market friendly, emphasizing efficiency, encouraging privatization, and looks to achieve win-win solutions that combine economic and environmental goals, at the expense of social justice. It individualizes responsibility, scapegoats consumers, excludes or exploits the poor, and enables corporations to elide responsibility and the state to escape accountability. It makes us feel good for doing our bit, comforting us with clean and sanitized spaces, all the while distracting us from the more fundamental contradictions of consumer capitalism. Bengaluru's communal sustainability proponents started shackled by this neoliberal thinking, only to be challenged by the inclusion of waste pickers in the zero-waste movement, who through their work and knowledge claims are expanding the conversation to include questions of sustainable livelihoods and justice.

In this concluding chapter, I take inspiration from the ways in which waste pickers and their representatives have engaged in a concerted program of popular education to explore transformative potential in the cracks of neoliberal sustainability. The presence of informal workers engaged in the metabolic flows of the city changes how sustainability manifests in Bengaluru. The waste picker, as the laboring body who engages in intimate contact with materials termed waste, garbage, scrap, or *maal*, knows the materialities and temporalities of waste in a way that average middle-class Bengalureans or bureaucratic functionaries do not. Waste pickers deploy their embodied knowledge through their work, documenting the reality of waste streams, as opposed to illusions of them as composed mostly of recyclable materials.

Waste-picker organizations, through their savvy use of social media, culture jamming, and coalition-building with the middle-class-led zero-waste movement (ZWM), have incrementally moved the city closer to some actions that target the political economy of waste, beyond disposal. These actions include a statewide ban on single-use plastics and a redesign of the waste collection system to restrict the power of garbage contractors. Waste picker-entrepreneurs like Deepak have become highly visible voices for change in the city. Yet, these gains are provisional and contingent. The commodifying and co-opting forces of global capitalism threaten to subsume place-based experiments into a corporate green-growth agenda. National campaigns prioritizing "smart cities," visual cleanliness, and privatization continue to undercut the power of labor. The progressive potential of place-based collective action is compromised by the denial of casteism, which locks most waste pickers into subordinate roles in recycling infrastructures.

Households and informal workers are the two groups most frequently scapegoated in the blame game of who is producing plastic pollution. When these two constituencies come together in organizing and advocacy, the state and corporations face more pressure and accountability. In Bengaluru, a move away from privatization, spectacular infrastructures, and corporate volunteerism toward public investment in coproduced and inclusive infrastructures shows us that unlikely coalitions can re-politicize the waste crisis. But if the urban sustainability agenda is to go beyond cleaning up or recycling, it must be led by diverse coalitions that both confront the throwaway cultures normalized by consumer capitalism and hold the state accountable to its most vulnerable constituencies. Most importantly, it must resist the seductions of growth, efficiency, and aesthetics to recover the urban commons from the detritus of a casteist society.

Engaged Universals for Thinking Sustainability

A key contribution of this book is to situate an analysis of neoliberalizing processes in relation to urban sustainability. How do neoliberal logics forwarded by the state and capital contour what it means to be a good environmental actor in the city? How do these logics constrain or open opportunities for participation by different groups? What are the sites, forms, and expressions of environmental action? And what sorts of social and material sustainability infrastructures emerge from the interactions of ideas, peoples, and flows? In answering these questions, this narrative resists seeing neoliberalism as a monolithic juggernaut refashioning hapless objects and subjects. Rather, I highlight how neoliberal logics are skillfully appropriated by diverse constituencies to make claims for themselves and for urban environments.

I offer up several concepts that I hope will function as engaged universals, ideas emerging from and grounded in a particular locality, but that aspire to travel and help make sense of other places and dramas (Tsing 2011, 8). It was by looking side by side at a middle-class spandex-clad cyclist and a *dhobi* on his bicycle that I began to think about the aesthetic and discursive biases of so-called sustainable consumption. I reworked the term *performative environmentalism* to name these biases. Now I see performative environmentalism everywhere I look, and especially in my home in a global hub of the "creative classes," the San Francisco Bay Area. Engaged universals are by definition provisional—they change and are changed by contact with other spaces. In Oakland, California, where I live, other axes of history, identity, and difference (race and indigeneity, for instance) matter to making sense of how performative environmentalism functions as a tool silencing communities who do not fit into the aesthetic and discursive registers of elite environmentalism.

Performative environmentalism is routine, habitual, and mundane. It is now woven into the very fabric of the everyday lives of the global consuming classes. Its everydayness makes it harder to question, problematize, and dismantle, as does the fact that most environmental scholars, including myself, are personally implicated in reproducing it. But an honest conversation about its limits is urgent because meeting climate targets requires absolute reductions in consumption (Creutzig et al. 2018; Dubois et al. 2019). In a world of mass poverty, this is possible only through degrowth and economic redistribution (Hickel 2020). While it might be tempting to declare that the dilution of environmentalism is purely a product of top-down efforts, I speculate that the everyday legitimating structures of performative environmentalism that reify the existing social order *also* perpetuate Band-Aid solutions. My argument suggests that an honest examination of "feel-good" environmental agendas practiced and promoted by the elites and middle classes—across the world—is a necessary step toward exploring alternative environmental possibilities.

Similarly, Bengaluru is not unique in its turn to community as a site for enacting sustainability. However, in Bengaluru you can clearly see that community is hierarchical. Caste, class, and gender divisions are in your face. You would have to be skillful at delusion to ignore them.[2] Looking from Bengaluru helped me name *communal sustainability* as a contradictory object, both empowering women and workers while responsibilizing them for social reproduction in a manner that intensified gendered, classed, and casted hierarchies. Employing *intersectional theories of social reproduction* to characterize sustainability initiatives can reveal community hierarchies and structures in finer detail than other approaches, I contend, and this lens can be applied across diverse cases and geographies. I also saw how the turn to individual responsibility and community participation in environmental action created openings for social groups who had been comprehensively excluded from participation in urban reform.

Groups representing waste pickers created a new mode of *entrepreneurial environmentalism* that put a spin on the environmentalism of the poor—a term usually used to describe the struggles of working-class, rural peasants and forest-dependent communities in the Global South against development and infrastructural projects that threaten their livelihoods and survival in the moment—to instead focus on how their livelihoods sustain urban environments. The metabolic dependencies on manual environmental labor embedded within communal sustainability created a new avenue for claims-making. In cities like Bengaluru that prioritize livability, aesthetics, and economic competitiveness as markers of success and sustainability, speaking about utility and usefulness is a more compelling entryway than rights-based environmental claims. Fashioning themselves as ecological actors also enabled waste-picker organizations to find common cause and form coalitions with middle-class environmentalists.

These coalitions in turn forwarded decentralized, coproduced infrastructural forms. The neoliberal state here again saw a convenient opportunity to further devolve infrastructural labor onto people through *DIY infrastructures*. DWCCs lock middle-class women and waste pickers into functioning as social infrastructure, keeping the city clean and green in a manner that is consistent with the state's goals of municipal de-risking and divestment. Geographer Rosalind Fredericks (2018) astutely points out that this technique of devolving infrastructural labor onto community is a key way the state seeks to divert and mute the activist capacities of urban residents. Making ordinary urban residents responsible for functions that would have been in the ambit of public control and responsibility gives them less time to make trouble. But it also makes the city more vulnerable to these people and social groups. Contradictions galore.

Neoliberal spaces, techniques, and subjects have multiple and contradictory aspects (Larner 2003). For instance, neoliberal reforms privatizing municipal waste collection resulted in a private contractor system that is universally loathed in Bengaluru. Middle-class zero-waste activists came to oppose garbage contractors because they found they made no effort to segregate waste or promote recycling. Labor movements opposed them because contractors subject *pourakarmikas* to inhumane working conditions, disrespect, and wage theft.[3] For Bengalureans unmoved by environmental or labor issues, contractors are still worthy of censure because they are not effective at keeping waste off the streets. The infrastructural breakdown provoked by privatization opened the way for experimentation with decentralized waste management systems.

Similarly, neoliberal discourses valorize the "consumer-citizen" to marginalize other social groups. The neoliberal state utilizes community participation as a softer, kinder mode of disciplining its subjects (Peck 2010). However, in the process, these discourses and apparatuses of participation also create openings for ordinary people to do politics. In many cases, bottom-up environmental mobilization dovetails with governmentalizing agendas (Roy 2009a). Surprisingly in Bengaluru, global connections and aspirations, as well as the waste-picker movement, meant that the ZWM was not in total accordance with governmental visions. Indeed, outcomes are not predetermined. Infrastructures of social reproduction can become spaces of contest and struggle.

In Bengaluru, infrastructural breakdown combined with a discursive emphasis on behavior change, consumer-citizenship, decentralization, and low-cost solutions opened room for some waste pickers to make themselves essential to middle-class schemes. It gave them a platform to make claims to recognition. These political claims challenge how unwaged, informal work and workers are rendered essential but disposable in capitalist economies. Formal enterprises, the state, and capital appropriate value from informal workers, but deny this metabolic dependence. Communal sustainability cracked open a door by forcing acknowledgment of the environmental and economic value of waste pickers through occupational identity cards and inclusion arrangements. It became the terrain where people who previously thought they had nothing in common—middle-class women and (mostly female) waste pickers—formed a cross-class coalition to push back against a male-dominated, technocratic, and centralized urban agenda. I contend that such cross-class coalitions are fundamental to recovering sustainability from its techno-managerial shackles, a point I elaborate on in the next section.

But the commodifying and co-opting forces of capitalism are a specter on the horizon. The effectiveness of community-education programs in changing household behavior, and waste pickers' success in recovering value from waste, has attracted a new breed of dispossessing actor. Start-ups and big business now see opportunities to turn municipal solid waste into a profit frontier. Segregated waste streams make it easier for companies to capture value from waste materials. Private players seek to retain waste pickers as a source of cheap manual labor and in ways that reduce their autonomy. Market-based circular economy initiatives concentrate power in the hands of polluters and investors and privilege new start-ups over informal organizations. The result is what I term *accumulation by inclusion*, where waste-picker inclusion becomes primarily about securing the conditions of growth and profit-making, leading to uneven outcomes for informal workers.

Looking at inclusion arrangements in Bengaluru shows us that cultural ideas about racialized others (as expendable, unworthy, devalued), explanations about the root causes of poverty (as a product of individual failure), and assumptions about how best to solve problems (through markets, efficiency, technological, and managerial innovation) influence how well-meaning sustainable development interventions are conceptualized and enacted. Unspoken but omnipresent ideologies frame who is worthy or unworthy and contour how waste pickers are engaged. The implicit preference for English-speaking, dominant-caste men talking tech privileges these voices over waste pickers who have toiled for years to recover resources. It reveals how environmental legitimacy and legibility are ontological, still predicated on caste/class position and transnational cachet, rather than on what one actually does. While waste pickers might have used performative environmentalism's aesthetic biases to make their claims, they are still not recognized as having ecological legitimacy in the same way the dominant-caste middle classes are. This is because of the unspoken ways in which casteism and, more broadly, racialization still hold sway in how informal workers are engaged in sustainable development.

As Assa Doron and Robin Jeffrey point out (2018), India's cheap labor is a magnet for trash from all over the world. It could now become a magnet for plastic producers and brands seeking to discharge their environmental commitments, at cheapest cost. But waste labor is not inherently cheap. It is cheapened through economic and cultural processes—such as colonial appropriation that produces huge wealth inequalities between nations and the social relations of racial/caste capitalism that devalue nonwhite/Dalit bodies to cut costs. Those forces shape the outcomes of well-meaning sustainability initiatives as well. The logics of racial capitalism can infiltrate environmental mobilizations and sustainable development practice in subtle and insidious ways, resulting in inclusion becoming a tool for capital accumulation in place of social justice. This is a warning to "inclusive circular economy" proponents and practitioners in particular—your preoccupations with inclusion as an antidote to injustice might lead to the very outcomes you seek to avoid.

Radical Incrementalism from Everyday Inclusion

Can the democratizing impulses of communal sustainability be recovered from casteism, "win-win" fallacies, and corporate capture? In asking this hopeful question at the end of this book, I am taking inspiration from feminist economic geographers who have called on scholars of poverty and difference to explore the ontology of a politics of possibility (Gibson-Graham

2006, 2008; MacGregor 2021b). The stories we tell as researchers can present the current situation as one that is unchangeable, a simple story of oppression repackaged. Or we can share humble narratives that leave room open for diverse futures. In chapter 5, I showed how the drive to include waste pickers as entrepreneurs articulates with a global circular economy agenda to perpetuate the conditions of growth and profit-making in the face of compounding ecological crises. I explain that inclusion becomes a tool for accumulation because inclusion efforts are individualizing, prioritizing profit over social outcomes, and because casteism goes unacknowledged as a disciplining force.

But inclusion is everyday and lived. It is a practice, not an outcome. The material practices of negotiation that take place within zero-waste movements and infrastructures can become the stage for citizenship struggles (Fredericks 2018). To make sense of material and discursive practices of negotiations as connected to and potentially generative of a broader transformation of social relations in the city, I turn to Edgar Pietrese's frame of radical incrementalism. Writing from situated understandings of African urbanism, Pietrese and other scholars of situated urban political ecology (SUPE) argue for an analytical framework that, instead of seeing power as residing in and enacted solely through social structures (i.e., through neoliberalism, capitalism, and so on), understands power as distributed and situated. SUPE examines city-making through everyday practices of power-laden negotiation among diverse groups. While negotiations are contoured by structural constraints, their outcomes are not determined by them. Situated analyses explore whether power can be reclaimed in incremental steps through recursive empowerment. Scholars of situated urban political ecology Mary Lawhon, Henrik Ernstson, and Jonathan Silver explain (2014, 511),

> The examination of diverse forms of power opens up spaces for reclaiming power in incremental steps. To be effective, such steps must be part of a larger project of recursive empowerment, for these actions "only begin to matter if one can effectively institutionalize such efforts" (Pieterse 2008, 131). This reframing allows for a reinterpretation of individualist practices, including the "quiet encroachment" of the urban poor (Bayat 2000), and the "performance of citizenship" (Scott 1985). At the same time, it calls for a critical consideration of when, how and under what circumstances such actions can be drawn into wider processes of resistance and empowerment. The process of turning these everyday moments into a radical incrementalism that supports recursive empowerment, as theorized by Pieterse (2008), is critical for understanding spaces of possibility and hope that

can multiply instead of evaporate or be placed within a centralized "state." Radical incrementalism is thus a situated, unfolding process which differs over time and across space.

Radical incrementalism and SUPE offer an important corrective to sustainability studies that remains dominated by epistemologies of the Global North. Sustainability studies encompass a tendency to imagine that it is possible to enact social transformation without engaging in transformational politics (Sovacool and Dunlap 2022). For instance, scholars have pointed out the need to make planning processes more inclusive of diverse perspectives (McLaren and Agyeman 2015). Others have called for environmental collective action that unites government, local citizens, researchers, NGOs, and business to make effective progress toward sustainable ecological and environmental planning (Nagendra 2016). The imagined processes of participation are citizens' assemblies or participatory forums, convened and supported by the state.

However, these ideas of polite, state-sanctioned deliberation ignore how culture and identity mediate participation in collective action (Mudliar and Koontz 2018). Official invited spaces of participation can deliberately exclude and silence dissenting voices (Ellis 2011; Miraftab and Wills 2005). Talk-centric deliberative fora wherein actors debate the merits of different possibilities ignore or undermine questions of identity and representation (White 2019). Caste, religion, and class operate as regimes of silencing and humiliation in Indian cities. Thus, the idealized version of a stakeholder forum in which all voices are heard and decisions that balance social, environmental, and quality of life concerns are reached through consensus has limited precedent in the Indian city. Indeed, when forums are convened, they are exclusive, either for elite and propertied classes or for laboring classes and the poor. The agendas of these groups rarely mix. When they do, the concerns of the poor and working classes are rendered less important and less valid, both implicitly and explicitly.

Recognizing these modes of silencing in urban environmental politics and lacking the numbers necessary to hold the state's attention through mass movements, waste pickers pursued a mode of engagement and negotiation that is embodied, accomplished through work with waste and representing that work back to the city. As we saw in chapter 4, waste-picker organizations, through savvy use of traditional and social media, reflect back to the city the metabolic flows of materials that zero-waste infrastructures are supposed to hide. Through this knowledge production and infrastructural negotiation, they sought to improve their conditions of life and work. I contend that this popular pedagogy also defamiliarized elite and middle-class understanding about the core drivers of unsustainability—debunking the simplistic, commonsensical mythologies forwarded by the anti-politics machine of sustainable development (Ferguson 1994).

Zero-waste activists thought they could fix Bengaluru's garbage woes by educating their neighbors, convening communities of practice, designing communal infrastructures, and unleashing market forces. It was waste pickers who mobilized their knowledge of waste materials and recyclability to show zero-waste activists that the problem was not just behavioral or infrastructural, but more fundamental. Careful documentation of dry waste flows busted the myth that the materials flowing through Bengaluru's households and commercial establishments were largely recyclable. Parisara Tanda organized DWCC visits for middle-class citizens on World Environment Day to demonstrate that, even with proper collection and sorting, the sheer volume of materials discarded in the city every day, much of which is toxic and hazardous, overwhelmed systems. It conducted plastic brand audits, recruiting middle-class environmentalists to join them in the activity. These national and international audits were conceived to highlight the culpability of plastic producers and brands for plastic pollution. Leaders in the waste-picking community like Deepak, Imran, and Shrimati conduct training sessions for college students, middle-class citizens, and other stakeholders, discussing the ongoing casteism and communalism that waste workers face. These educational tactics have been also employed by waste-picker organizations in Brazil and Argentina to build awareness among the general public about the socio-environmental aspects of waste management (Gutberlet et al. 2021).

The conditions of waste-picker work, material realities, and their representations have shifted the sustainability agenda in Bengaluru. What started as a movement to keep waste off the street evolved, first, to include a more ecological understanding of the issue, and eventually included topics that the middle-class zero-waste movement would have never thought about alone: occupational health, housing, and social security for waste pickers, reversing privatization of waste management, and a serious discussion about holding the producers of plastic and nonrecyclable materials accountable. In 2018, the Karnataka state government imposed a ban on manufacturing, storing, distributing, and using single-use plastics, partially because of this activism. María José Zapata Campos and coauthors document how waste-picker networks in Africa and Latin America produce discourses and rationalities of social inclusion and environmental sustainability, and engage in coalition-building, as political strategies. Through this they challenge and expand environmental governance possibilities (Zapata Campos et al. 2021). In similar ways, grassroots environmental stewardship by some waste pickers is demythologizing and defamiliarizing the common explanations for urban dysfunction and unsustainability. Not only do waste pickers play an important role in knowledge production promoting recycling and recovering more resources, but their knowledge also helped demonstrate that current systems of tackling pollution were fundamentally flawed and that the core problem lay further upstream. In unequal, elite-dominated cities, re-politicizing waste, pollution, and unsustainability is an incremental process of embodied communication.

Waste Activism as a "Contact Zone"

As environmental movements recognize their ecological legitimacy and expertise, some waste pickers have become public figures and thought leaders in the city. Waste picker-entrepreneurs like Imran and Shrimati are featured on national and international webinars about waste management and climate action. They are quoted in the media. They participate in global climate summits. They make videos that are distributed on Twitter and Facebook. Leadership and advocacy activities also put waste pickers and middle-class zero-waste activists in increasing contact. I contend, hopefully, that cross-class coalitions in Bengaluru potentially provide opportunities for the kinds of spatial encounters in contact zones that can lead to transformative moments in how the middle classes politicize both unsustainability and poverty.

Feminist geographers Victoria Lawson and Sarah Elwood challenge researchers of poverty and class to search for the "ontologies of a politics of possibility." They describe this outlook as "an ontology of openness and curiosity about what is empirically present, but what often remains theoretically unseen: middle-class actors who engage with 'poor others' in ways that lead to shifts in normative understandings of poverty and enacted poverty politics" (Lawson and Elwood 2014, 210). Failing to look for instances where the dominant ways of explaining poverty and the poor are challenged might mean researchers discount moments that create new possibilities.

Uneven, distributed, and erratic contact zones are emblematic of the urban present in Indian cities. The proliferation of survival jobs such as waste picking, domestic servitude, and food delivery led by rapidly rising middle-class affluence put the middle classes in contact with socioeconomically and caste-marginalized groups on a regular basis. This contact has always involved contradictory aspects. Moreover, as Indian cities become more unequal and the state steps back further from any commitment to supporting its poorest constituencies, the underclass becomes more dependent on the patronage of the affluent to survive. New forms of patron-client relations fill in, imperfectly, for the postdevelopment state (Gidwani and Reddy 2011; Kornberg 2020). As previous chapters have shown, expunging waste from middle-class homes frequently involves domestic help who serve as the conduits of refuse from the home to the street or the dump.

The zero-waste movement also implicated the urban underclass, seeking to discipline maids, drivers, and sweepers into changed behaviors. The contact that happens around waste and even zero waste represents this form of patron-client relationship. However, the contact zones convened via joint activism are different. What makes them different is the storytelling that waste pickers and their advocates do in these spaces, a storytelling that has performative capacity and persuasive power. Although contact zones emerge from pragmatic motivations, waste-picker knowledge production can turn these into zones of transformative encounters, where middle-class actors are forced to reexamine their understandings of waste, waste pickers, and social systems more broadly.

In 2013, when I was supporting advocacy efforts led by TCGF and other groups, I participated in the creation of one such contact zone, where members of the middle classes encountered and engaged with waste pickers in counter-hegemonic ways. I helped organize a waste management awareness event. This event, a collaboration between the local elected representative's office and a collection of neighborhood associations, was envisioned as a launch point for a zero-waste management plan and was attended by around two hundred residents. As part of the event, the organizing committee decided to convene a series of panel discussions on zero-waste implementation. When choosing panelists, we proposed inviting a waste picker to participate in the discussions. With Parisara Tanda's help, the event was attended by Santosh, a waste picker who had just begun to operate one of Bengaluru's newly commissioned DWCCs. Santosh, who had working knowledge of three languages, including some English, participated in the panel discussion as an expert on recycling. The other members of the panel included a Brahmin woman who had launched a waste management scheme in her luxury apartment complex, an elected representative, a doctor who had expertise in public health, and a composting expert. During the session, Santosh explained, "We, waste pickers are dependent on you, because if you stop us from working, we will starve. But, you are also dependent on us, because if we don't do our work there will be even more waste on the street. So we have to find a way to work together to make everything better for all of us."

The story of Savitha, a fifty-year-old housewife living in northcentral Bengaluru who co-organized this event, illustrates the transformative impact of interaction in contact zones. In a 2017 interview, Savitha confessed to me that she never thought she would be so involved with waste pickers in the city. Her engagement in the ZWM started in the usual way; she was a member of her local residents' association for several years and an influential voice in neighborhood politics pushing for clean streets. During her engagement with local civic campaigns, she became interested in zero-waste management as a potential solution to ongoing garbage issues in her neighborhood. She joined TCGF, met activists who organize informal workers, and began to engage with some waste pickers. These personal engagements changed her views on waste pickers. She began to see that they shared similar goals—to live in a safe, clean city. Savitha is not alone. Several middle-class waste activists have become supportive of waste pickers because of their in-person encounters. Sheetal hosts waste-picker training sessions in her home. Her husband is not a fan, but she has decided to go ahead without his approval. Waste pickers have also at times been able to rely on middle class support to resist intimidation from garbage contractors. As one labor activist put it, "Green conscious middle classes helped us fight the garbage mafia." Now, a sizable contingent of environmentally engaged middle classes ally themselves with waste-picker goals and rights and vouch for waste pickers in public fora.

Yet it is essential to emphasize that these moments of transformative possibility are just that, ephemeral moments. Assorted transformative moments in contact zones have not eliminated the hierarchies and exclusions of communal sustainability. Many middle-class "environmentalists" continue to malign the work of waste pickers. Alongside labor-friendly moves to regularize sanitation work are a series of antipoor measures implemented by the municipality in the past three years—often with the encouragement of some middle-class actors. The BBMP has paused the process of issuing occupational identity cards to Bengaluru's waste pickers. Hefty fines for littering enforced by ex-military marshals, and a single-use plastic ban that does not include any provisions to help informal vendors transition to using other materials, could harm the city's poor and working classes in unexpected ways. Entrepreneur inclusion models have also segmented informal waste workers, with many migrant waste pickers remaining in precarious positions (AICCTU 2021). And while there are attempts to draw individualist practices of waste picker advocacy and leadership into wider processes of resistance and empowerment, the lack of a strong mass movement leaves waste pickers vulnerable to changing policy priorities.

Recognizing these vulnerabilities and the need for a collective approach, on March 1, 2022, which commemorates Global Waste Pickers' Day, waste pickers and small scrap dealers launched a new statewide member-based organization, Thyajya Shramika Sangatane (TSSK), that would unite around fifteen thousand informal waste workers across thirteen cities and villages in Karnataka. Leveraging its numbers and statewide scope, TSSK is now pushing the Karnataka state government to restart issuing identity cards and provide waste pickers with health insurance, disability insurance, and pension fund accounts. This story is still unfolding. While it remains to be seen if waste pickers can build the collective might needed to reverse privatization, break caste, and improve their lives, sustainability and environmentalism have served as necessary and unavoidable terrain for their struggle in Bengaluru thus far. And this struggle is not theirs alone.

Solidarity and a sense of mutual vulnerability between progressive, environmentally engaged middle classes, and informal workers is even more important today. The postdevelopment Indian state is increasingly an authoritarian one, empowered by a large section of society that ascribes to its Brahminical Hindu Nationalist sentiments (Ranganathan 2021, Reddy 2021). Waste pickers, as poor and largely Dalit and Muslim groups, are especially vulnerable at this time. Conversely, when middle-class climate change activist Disha Ravi was arrested and charged with anti-national activity, environmental and social progressives received a terrible wake-up call (Kashwan 2023). Their class and caste privilege no longer offered protection from the state's repressive tendencies. Authentic collaboration with working class labor movements and anti-caste movements will be necessary to sustain environmental activism and advance sustainability in the country.

From Performative to Reparative Environmentalisms

We are at the end of this story. Where do we go from here? Even though waste pickers are now recognized as environmental heroes, as we see from this story, "recognition and intent are grossly insufficient for changing systems of discard, or even to study them ethically" (Liboiron and Lepawsky 2022, 120). As theorists of decoloniality contend, gestures of inclusion can function to maintain an unequal social order (Glenn 2015; Maldonado-Torres 2011; Mignolo 2007). Bengaluru's local waste and pollution crises are connected to and inseparable from a global crisis of plastic overaccumulation and climate change. They are also locally complex, challenged by the sheer magnitude, density, and frequency of waste generation, as well as the casteism that has stymied the development of modern sanitation infrastructures. To overcome false solutions that reproduce the exploitation of oppressed communities and fail to tackle the core drivers of ecological destruction, we need a fundamental shift in thinking and action beyond inclusion.

The onus to create systems that "discard well" is not just on waste pickers, but also on scholars and activists concerned about sustainability and ecological health.[4] What are the steps we can take? In closing this book, I want to offer some ideas, big picture and specific policy proposals, to ally with and strengthen the radical incrementalism that waste pickers are attempting to enact in Bengaluru and beyond. These ideas are for those of us who see ourselves as committed to the broader project of ecological health, environmental justice, and equitable, sustainable well-being. I am speaking to you, my fellow well-meaning researcher, activist, engineer, or policymaker.

Reclaiming environmental questions from technocratic, managerial, and economistic capture is a necessary first step. Thinking relationally, historically, and spatially reveals how environmental pollution is produced and enacted through colonial land relations and (racial) capitalism's cheapening strategies that turn people and places into sinks (Liboiron 2021; Patel and Moore 2018). Dismantling these structures of oppression will be a long, hard road ahead, one that cannot be traversed by middle-class environmentalists or labor groups alone. Rather, confronting the interlinked global-local environmental crises of climate change and environmental pollution will require transformative coalitions led by frontline communities.

Waste pickers and other waste workers know the most about the material natures of waste. Leveraging this knowledge and building coalitions, waste pickers recursively empowered themselves. Through their advocacy and embodied activism, they illuminated the fallacies of neoliberal sustainability. Waste pickers showed the ZWM that efficiency and entrepreneurism was not going to get Bengaluru out of its garbage mess. They demythicized "clean as green," teaching middle-class environmentalists that if the sustainability agenda is to succeed in achieving a healthy city, it must confront more centrally the throwaway cultures normalized by consumer capitalism. Waste pickers are doing environmental activism while trying to survive in a world that treats them as disposable. They are constantly negotiating eviscerating forces that repackage their exploitation with a green sheen.

Communal sustainability, as emerging from and practiced by middle-class zero-waste warriors, politicizes the waste crisis by connecting it to urban development and consumerism, but it does not question caste, and thus struggles to emerge as a transformative coalition centering justice and equity. Communal hierarchies and differences (along lines of race, caste, religion, ethnicity) undermine collective and communitarian routes to social transformation, as do individualizing narratives that reduce people to consumers and entrepreneurs. The reliance on caste-determined environmental labor to produce sustainable cities ultimately serves the ends of capital accumulation, undermining environmental and social goals. Indeed, what waste-picker coalitions are trying to do world over is resist the colonial, casteist, and racist ways in which the powerful hide and deny the oppression that enables capital accumulation. Instead, they demand that *they*, and not just their exploitable labor, be seen as indispensable to our collective futures.

Looking at this case study in Bengaluru and zooming out from it reminds us that our systems of production and consumption do not work for most people in the world. Whether it is waste pickers materially deprived of basic dignity and forced to sort through refuse to make life, or the middle-class mother in Bengaluru worried about her child's fourth bout of malaria,[5] waste reveals a broken economic system. Recognizing the interlinked nature of ecological violence and human suffering across diverse positionalities can help us see our mutual vulnerability. Waste flows form the material bonds of our mutual vulnerability. Our mutual vulnerability is what we have in common.

Recognizing mutual vulnerability and racial and socioecological indispensability means rejecting the ideas of dominion that legitimize the dehumanization and oppression of others (Pellow 2016). In India, this means the annihilation of caste. Quite simply, there can be no *just sustainability* as long as environmental initiatives ignore, condone, or reproduce casteism. Scholarship on Indian cities and environmentalism is finally beginning to include caste within its analytic, thanks largely to the efforts of Dalit scholars and activists (Aiyadurai and Ingole 2021; M. Sharma 2017b). Rejecting casteism involves recognizing the reality that salubrious environments and economic gains for dominant-caste, middle-class, and elite groups in India has always come at the cost of oppressed castes. Thinking about casteism as an expression and form of racialized oppression connects the struggles of Bengaluru's waste pickers to the struggles of racialized and othered communities all over the world (Ranganathan 2021).

Anti-casteism, like antiracism, is not simply an orientation but a practice. Enacting an anti-casteist, justice-oriented sustainability will require not just new ways of thinking but also new ways of doing and being in the world. And we cannot wait to abolish caste to reform dehumanizing waste management systems either (Swaroop and Lee 2021). These actions go together. Achieving salubrious environments without caste oppression will require changes from everyone—households, municipal governments, businesses, and the state. And while no systemic solution to Bengaluru's waste problem is likely to be perfect, and there is no single and universal good that can be achieved, one that combines an ecological ethic with an anti-casteist practice is more likely to be a system that "discards well"— especially if we consistently prioritize the well-being of communities who have thus far borne the brunt of a wasteful city.

Converging social and ecological conditions present an opportunity for changing this broken system. The COVID-19 pandemic had a devastating impact on waste workers across the world. It cut waste pickers off from the source of their livelihoods and exposed them to health risks while also making it harder to access aid and community resources (Chandran 2021; Hartmann, Hegel, and Boampong 2022). Women waste pickers were particularly vulnerable to hunger and increased precarity (Wittmer, Srinivasan, and Qureshi 2020). Sanitation workers too toiled amid great hazards—three *pourakarmikas* in Bengaluru lost their lives as they worked on the front lines without any protective equipment. Yet, COVID-19 also reminded elites and the middle classes of their fundamental dependence on sanitation workers and waste pickers, and of their own vulnerability. As Kanthi Swaroop and Joel Lee reflect (2021, 39), the "emergence and growth of a discourse in which sanitation workers are guardians of societal well-being rather than objects of caste contempt" has in some cases improved the material conditions of their work. Meanwhile, pandemic "risk society" increased the use of disposable plastics, bringing new urgency to the task of tackling plastic pollution (Silva et al. 2021). The conjoining of labor and environmental crisis offers the opportunity to see these oft-separated issues as inexorably linked.

In closing this book, I want to argue for an alternative agenda of *reparation by inclusion*, which rejects the dehumanization of informal workers and the seductions of performative environmentalism. An ethic and practice of reparation seeks to reimagine and recreate socio-ecological relations from a full acknowledgment of the injustices of the past as they live into the present. I advocate reparation over justice because dominant, liberal notions of justice center the individual, foreclosing consideration of histories of harm and denying the need for collective redress. Reparation is less open for capture by neoliberal capitalism's individualizing, commodifying, and depoliticizing forces.

My thought-partner Jennifer Tucker and I argue in a 2020 perspective piece that the frame of reparation offers a corrective to the dominant logics of legalistic and deficit-based interventions in informal economies (Tucker and Anantharaman 2020, 295).

> Decentering formalization, we advocate for reparation as an ethic to orient the actions of development practitioners and policymakers. Our debt here is to the Black radical tradition (hooks 1996), scholars and activists proposing collective redress for the unspeakable violence and thefts of slavery, legacies that live into the present (Martinez-Alier 2016; Wiedmann et al. 2020; Liboiron 2021; Hickel 2017). Following W. E. B. Du Bois, the ethic of reparation activates memory against the forces of willful forgetting that deny history and deep relationality (Du Bois 1935). We are inspired by the Black radical tradition's expansive, future-oriented political imagination and its call to remake economic and social relations from the roots up (West and hooks 2016). "Reparation ecologies" add an imperative to heal the false Nature/Society divide, locating socio-economic relationships within

living ecologies. Both lines of thinking emphasize redistribution: of resources, land, work and the labors of care. While formalization targets workers or the economies that sustain them for reform, the field of action promoted by reparation is much broader, including the forces producing inequality and environmental harm. Reparation acts horizontally, centering communities most harmed by fomenting worker power, repairing historic injustices, and redistributing social power and resources to the grassroots. Reparation can also help us acknowledge the long history of assent to exploitation that structures scholarship and practice (Roy 2017).

A *reparation by inclusion* agenda leverages sustainability and circular economy to demand more material and intellectual investment into informal economies, as well as access to more social entitlements for informal waste pickers. It starts with a claim that informal economies have long subsidized the conditions of capital accumulation, often through (self) exploitation, and that any sustainability initiative will have to redress existing injustice through grassroots leadership. When this investment is made from an acknowledgment of past injustices, it avoids a deficit-based agenda of disciplining waste pickers. At UNEA assemblies and global climate summits, grassroots waste-picker organizations are already insisting that sustainability practitioners go beyond economic and ecological rationality: they are demanding full and meaningful participation in national and international policymaking, fair wages and social security, and systematic plans to phase out toxic substances in plastics. By delinking labor, income, and development, and connecting social reform with ecological healing, these reparative policies foster more just ways of organizing work, time, and life.

Reparation by inclusion is impossible within a sustainability regime that justifies every action based on narrow cost-benefit analysis, efficiency metrics, and aesthetic preoccupations. Such a sustainability regime is fated to reproduce injustice. Instead, sustainability can become a terrain through which ecological and social harm can be repaired *if* there is a recognition of mutual vulnerability, a refusal to participate in dehumanization, and a reparation of power and resources. The circular economy's emphasis on sharing, reuse, and regeneration holds promise for ameliorating both ecological harm and social inequity, but only if its interventions are designed with attention to power, scale, and difference (Liboiron and Lepawsky 2022, 132). This means redistributing power so informal workers are the ones developing EPR policies and determining the terms of integration. It means the redistribution of revenue toward waste pickers. It means using technology not to increase profit, but instead to stop exposing waste workers to social and material toxicity. It means investing in public infrastructure and in social schemes like care incomes and universal basic services to ensure a *just transition* for informal waste workers exiting this occupation.

Sustainability initiatives that are driven by an ethic and practice of reparation would redistribute resources and power toward oppressed communities and disentangle socio-ecological sustenance from corporate growth or accumulation goals. Building intersectional coalitions led by those most harmed by the status quo, which remake everyday life alongside doing politics, is the key to a more transformative sustainability. To move forward, privileged environmentalists, scholars, and activists must accept leadership from waste pickers, sanitation workers, and other frontline communities, and prioritize social reform and reparation over aesthetics or efficiency.

Notes

Acknowledgements

1. Manisha Ananthraman, "Is It Sustainable Consumption or Performative Environmentalism?"*Consumption and Society*, no. 1 (2022): 120–143.

2. Cindy Isenhour and Paul Roscoe, eds., "On Social Change, Status, and Sustainability Policy," *Consumption, Status, and Sustainability: Ecological and Anthropological Perspectives* (UK: Cambridge University Press 2021): 324.

Introduction

1. The tech-industrial park was a SEZ, or special economic zone. A special economic zone is a geographical region designated by the central and state government wherein laws and regulations pertaining to taxes, labor, or export are more liberal than in other parts of the country. Creating SEZs is a strategy to attract investment, encouraged by structural adjustment policies, resulting in land dispossession in many cases (Levien 2011).

2. Yes, you read that right. The Indian Tobacco Corporation, now just known as ITC, is one of India's largest manufacturers of fast-moving consumer goods. It ran this recycling program as a corporate social responsibility initiative.

3. For example, discarded milk packets sell for 35 rupees (0.5 USD)/kg versus 2 rupees (.03 USD)/kg for unsegregated dry waste. Fundamentally, the value of waste goes up with sorting and aggregation.

4. The names of all individuals and civil society organizations whom I engaged in this study have been anonymized with pseudonyms. All interviews were conducted in confidentiality, and the names of interviewees are withheld by mutual agreement.

5. A public interest litigation is a legal tool that enables ordinary citizens to approach the court on any matter of "public interest." Although initially conceived to enable those in socially and economically disadvantaged positions to access justice through

the higher courts, the PIL has since become a key tool through which middle-class individuals and groups seek to influence public policy (Bhan 2009; Rajamani 2007). This PIL was filed in the High Court of Karnataka in June 2012 by two individuals, one of whom was informally associated with a middle-class waste management advocacy group. It petitioned for the implementation of the Municipal Solid Waste Management Rules 2000 (MSW 2000), a central directive that dictates how municipalities are required to manage solid waste. The Municipal Solid Waste Management Rules were issued by the Ministry of Environment and Forest, Government of India (GoI), in the year 2000 in response to a prior public interest litigation filed in the Supreme Court of India. Municipal Solid Waste Rules 2000 stipulated how urban local bodies in Tier I and Tier II cities should handle solid waste. They were framed by a special committee in response to a public interest litigation filed in the Supreme Court by Ms. Almitra Patel in 1996 (Almitra Patel vs. Union of India, WP 888 1996). The MSW 2000 rules promoted door-to-door collection of waste, segregation of waste, scientific landfilling, and increased community involvement in solid waste management.

6. Several newspaper articles and reports documented the conflicts at Mavallipura and Mandur (*Deccan Herald* 2013; Environment Support Group 2010; Ramani 2012; *Times of India* 2014). The Mavallipura site was finally shut in 2015, after years of concerted struggle by the Dalit Sangarsh Samithi and Environment Support Group (ESG). The site is yet to be fully cleaned up (EJOLT n.d.).

7. The garbage crisis was covered in local and global media (Beary 2012; Harris 2012).

8. Directions were issued by the court while hearing the PIL filed by ESG and ors. (WP No. 46523/2012) and other, related PILs. For more on these court proceedings and how they represent a departure from prior examples of judicial activism, see Lutringer (2017) and Lutringer and Randeria (2017).

9. The municipal solid waste management rules were revised and updated in 2016. The current rules place more emphasis on source segregation of waste. The High Court decisions in Bengaluru, as well as involvement of activist groups from the city in the national deliberations, had a significant impact on what was included and excluded (Luthra 2020).

10. A flagship initiative of the Bharatiya Janata Party government, the Swachh Bharat Abhiyan, or "Clean India Mission," aims to clean up India through various initiatives including ending open defecation and littering, increasing responsible waste disposal, and educating the population on the importance of cleanliness. Swachh Bharath and its local predecessor Swachha Bengaluru largely frame urban filth as a product of bad behavior as opposed to unchecked consumption growth, inadequate infrastructure, or inequities in service provision. Its techniques involve ranking cities by cleanliness, disciplining workers, privatizing municipal services through public-private partnerships, and pushing for large-scale, capital-intensive technologies like waste-to-energy plants.

11. The notion of just sustainabilities requires centering justice at the core of all sustainability struggles. It asserts that domination and oppression are incompatible with sustainability, and thus brings attention to the ways in which actions carried out in the name of sustainability either dismantle or reinforce systems of oppression (Agyeman 2013; Agyeman, Bullard, and Evans 2002; Agyeman and Evans 2003; Castán Broto and Westman 2019).

12. Several books and journal articles have chronicled the economic and sociological characteristics of the new middle classes and the transformations of the city alongside their emergence (H. Gupta 2019; Nisbett 2020; Pani, Radhakrishna, and Bhat 2010; Upadhya 2008; Upadhya and Vasavi 2008).

13. Around the world, ailing and disinvested waste management systems represent state dysfunction and failed promises. While mobilization around garbage often takes the form of strikes and protests targeting the state, and sometimes spiral into broader political demonstrations, as in Lebanon or Tunisia (Ball 2021; Yee and Saad 2019), the everyday nature of garbage, the way it surrounds you and impedes your life, means that it also sparks community mobilization at the neighborhood scale (Fredericks 2018, Stamatopoulou-Robbins 2019).

14. See Ram (2019). While this sounds like a big number, it is significantly less than cities in the United States and Europe. See Vergara and Tchobanoglous (2012). US and European cities simply do a better job of hiding and displacing their municipal solid waste through more extensive infrastructural arrangements. Bengaluru and other Indian cities also have the additional challenge of being very dense (Doron and Jeffrey 2018, 43).

15. See Doron and Jeffrey (2018), OECD (2007).

16. Here I am echoing Rob Nixon's classic text, *Slow Violence and the Environmentalism of the Poor* (Nixon 2011).

17. As analyzed in several recent studies (Cardullo and Kitchin 2019; Das 2020; B. Ghosh and Arora 2022).

18. The International Labour Organization (ILO) estimates that somewhere between 19 to 24 million people worldwide make a living as waste pickers (ILO 2018). In India, rough calculations estimate that its cities support about 4.5 million waste pickers (Doron and Jeffrey 2018, 189). However, reliable numbers on waste pickers are hard to obtain, and numbers fluctuate seasonally. Gleaners are a feature of North American and European cities, too (Bonatti and Gille 2019; Porras Bulla, Rendon, and Espluga Trenc 2021; Wittmer and Parizeau 2016). In many US cities, they are considered to be "stealing" value from the city, which sells its recyclables to subsidize the cost of running curbside recycling programs and generate profit. Curbside recycling programs in the United States emerged in the 1960s and 1970s, actively promoted by plastic producers and brands to circumvent more punitive

regulation. Geographer Max Liboiron names these "rigorous control" practices that contain pollution through "municipal curbside collection, industrial-scale recycling and highlight controlled and technical landfilling" (2021, 75) as emerging from a colonial perspective and colonial land relations in the illuminating and challenging book *Pollution Is Colonialism*.

19. Several studies have attempted to calculate the contributions of waste pickers to cleaning the city. A study by the International Labor Organization in Pune, India, estimated that scrap collectors salvaged 144 tons of recyclable scrap each day, saving the municipality millions of rupees (Chikarmane and Narayan 2005). Another study claims that every ton per day of recyclables collected by the informal sector saves cash-strapped urban local bodies 24,500 rupees per year and avoids the emission of 721 kg of carbon dioxide per year (Annepu 2012).

20. The term environmental privilege has been coined by scholars of critical environmental justice to indicate the flip side of environmental injustice (Park and Pellow 2011; Pulido 2015).

21. Scholars writing about South Asian environmentalisms have emphasized the importance of considering how social position and political-economic power contour the way urban environmental problems are framed and engaged. These approaches combine a Foucauldian lens of governmentality with a focus on social position (Doshi 2019; Rademacher 2011). Such an approach sees environmental subjectivities as context-dependent, in which context is socio-material infrastructure and governing regimes, equally influenced by how people name and make sense of their socio-material contexts, which are in turn influenced by both locally emergent and globally circulated ideas.

22. The apolitical and technocratic framing of sustainability has been analyzed and critiqued in several recent books (Castán Broto and Westman 2019; Hodson and Marvin 2014; Isenhour, McDonogh, and Checker 2014).

23. For feminist geographers Victoria Lawson and Sarah Elwood, contact zones are "mutable sites/moments of interaction in which differences are made explicit and can lead to new negotiations of identity, privilege, political responsibility and alliance." In other words, they are places of possibility (Lawson and Elwood 2014, 211).

24. Approaching debates on sustainability from a critical environmental justice perspective, Julie Sze and coauthors argue for a "situated sustainabilities" model as a way to explore how sustainability (research, policy, and practice) can avoid the trap of reinforcing the dominant ideologies that produce social injustice and environmental harm. Their framework connects environmental justice research and its commitment to praxis, racial justice, and examination of positionality to questions of sustainable development (Sze 2018).

25. Sociologists and political scientists have studied the illiberal and exclusive expression of civil society activism in India, emphasizing how the state encourages

elite and middle classes to participate in governmental community-participation schemes while making it harder for working-class groups to do so (Coelho and Venkat 2009; Ellis 2012; Ghertner 2011b; Kamath and Vijayabaskar 2014).

26. Sharad Lele and coauthors, in *Rethinking Environmentalism*, call for more collaboration among all academic disciplines in which societal-environmental interfaces are investigated, irrespective of ontological and epistemological stances (Lele et al. 2019).

27. Here I take my cues from the field of comparative environmental politics, which analyses how the differences in political systems contours who participates in policy-making, what types of policies emerge, and their environmental and social outcomes (Kashwan 2022; Steinberg and VanDeveer 2012).

28. As David Schlosberg (Schlosberg and Craven 2019), Francesco Forno (Forno and Graziano 2014), and other environmental politics scholars note, a growing number of environmental groups in the Global North are shifting their focus from political claims targeted at the state to instead engendering more sustainable practices in everyday life. These movements are sometimes branded as "sustainable materialism," as they often involve substantive, material interventions into everyday life infrastructures. Scholars disagree on whether these material interventions are political or apolitical, new or long-standing, and on whether they have transformative potential in terms of achieving just sustainabilities (M. Anantharaman et al. 2019; MacGregor 2021a).

29. Postpolitical, or more generally, the processes of depoliticization, refer to a situation in which the political, understood as a space of contestation and antagonistic engagement, is increasingly colonized by politics, understood as technocratic mechanisms and consensual procedures that operate within an unquestioned framework of representative democracy, free market economics, and cosmopolitan liberalism (Swyngedouw 2009). In postpolitics, political contradictions are reduced to policy problems to be managed by experts and legitimated through participatory processes in which the scope of possible outcomes is narrowly defined in advance. "The people," as a potentially disruptive political collective, is replaced by the population, the aggregated object of opinion polls, surveillance, and bio-political optimization. Citizens become consumers, and elections are framed as just another "choice" in which individuals privately select their preferred managers of the conditions of economic necessity (J. Wilson and Swyngedouw 2014, 6).

30. Community participation functions as a tool of "soft-neoliberalism," repackaging austerity and the absence of an ethical, social contract under feel-good tropes of citizenship and participation. For more, see Peck (2010), Fredericks (2018), and Miraftab (2004a).

31. As discussed in chapter 2, Bengaluru has a long tradition of middle-class civic involvement attributed to the city's larger middle-class population, self-association with modernity through presence of educational institutions and public-sector enterprises,

large diasporic population, and state schemes promoting middle-class involvement in urban governance (Gopakumar 2020; H. Gupta 2018; Upadhya 2017).

32. This reflects what Raka Ray and Seemin Qayuum describe as "cultures of servitude," characteristic of middle-class habitus (Ray and Qayuum 2009).

33. The word "communal" itself has this double meaning. Its positive usage refers to sharing, cooperation, and collective ways of provisioning everyday life. However, growing up in India at a time when Hindutva was on the ascent, I associated the word communal with chauvinism, riots, and violence. This double meaning applies to my analysis of communal sustainability as well; it is shared and cooperative but holds deep divisions within it. This analysis recognizes the messy and contradictory nature of community and participatory processes in urban development (Rigon and Castán Broto 2021).

34. "Environmentalism of the poor" was a term developed by Joan Martinez-Allier and others to identify and legitimize the livelihood struggles of agrarian and forest-dwelling communities in the Global South as forms of environmentalism (Martinez-Alier 2016). The term was developed to counter the then-dominant notion that environmentalism was a post-material concern—that people could only become environmentalists when their material needs were satisfied. Rather, Allier and others argued that environmental concerns are always deeply material concerns. In doing so, many more livelihood and land struggles became visible as environmental justice struggles. However, a broader criticism leveled at the framing of "environmentalism of the poor" is that it can lend itself to a form of eco-traditionalist discourse that essentializes those working in or with land as inherently ecological, and sometimes denying them the right to modernization. Further, the urban poor are less able to make ecological claims in this vein because nature is less visible in the city and their occupations look more extractive.

35. Political scientists studying Indian civil society have noted that members of the informal working class are largely excluded from participation in civic organizations. Slum-based collective action is a key mode through which some sections of the poor and working classes are able to make claims on the state (Coelho, Kamath, and Vijayabaskar 2020; Harriss 2006; Kamath and Vijayabaskar 2014).

36. Chapter 3 goes into this in more detail.

37. This tactic has been used by rural environmental struggles as well (Kashwan 2023).

38. This finding resonates with what Anne Rademacher found in Kathmandu, where, when riverbank slums were threatened in river restoration projects, their residents were fashioned as good ecological subjects to make resettlement claims (Rademacher 2011). However, while her Foucauldian analysis focuses on the formation of environmental subjectivities, mine also explores the types of social mobilizations and coalitions that these subjects engage in.

39. Studies of infrastructural citizenship have used water and waste infrastructures as key terrain on which to explore questions of political mobilization and justice (Anand 2017; Fredericks 2018; Moore 2012; Ranganathan 2014).

40. Citizen participation schemes and civil society involvement is almost exclusively restricted to propertied groups in Indian cities (Ellis 2011; Ghertner 2011b; A. Ghosh 2005).

41. Indeed, as I was finalizing this manuscript in 2022, and during my last field visit to Bengaluru, I learned that the BBMP, through the newly created Bengaluru Solid Waste Management Company (a parastatal body tasked with managing waste in the city), was threatening to revert to a policy in which door-to-door collection would be done by a single contractor, who would then drop off dry waste at the DWCC. If implemented, this would be a serious setback to waste pickers operating DWCCs (Menezes 2022b).

42. Waste pickers are more organized and involved in political spaces of negotiations now, and the environment is a key arena for their involvement (Chen, Carré, and Carré 2020; Dias and Samson 2016; Gutberlet 2021; Rosaldo 2019).

43. Political scientists and commons scholars have studied the effect of sociocultural heterogeneity on environmental collective action, finding that the effect of heterogeneity on encumbering or encouraging collective action is mediated by institutions (for a good summary, see Mudliar and Koontz 2018). These studies are an important corrective to the methodological individualism embedded in how Elinor Ostrom (1990) originally conceptualized collective action and community-based environmental management. As Kashwan et al. note, methodological individualism prevents many commons scholars from advancing a more nuanced exploration of how social, cultural, or political inequalities shape community mobilization and collective action for governing the commons (Kashwan et al. 2021). Other intellectual traditions, such as those emerging from Black Marxism or from Foucauldian approaches, offer a different perspective—they argue that identity formation and social difference is itself tied to specific institutional forms and the political ecologies associated with them. Theorists of racial capitalism, for instance, show that capitalism tended to "differentiate—to exaggerate regional, subcultural, and dialectical differences into 'racial' ones," to justify dispossession and colonial appropriation (Kelley 2017; Robinson 2000). Or, from a Foucauldian perspective, social locations are not abstract categories but are deeply enmeshed in and constituted through elite-based governing regimes (Agrawal 2005; Doshi 2019).

44. Critical theorists of environmental justice who incorporate theories of intersectionality (Crenshaw 2017; Kaijser and Kronsell 2014) explain that social location affects not only how different groups experience environments or express environmentalism, but also that how social identities are constructed and maintained to serve a status quo (such as colonial expropriation, capitalist exploitation) contours how

environmental issues are problematized and acted on, and to what ends (Amorim-Maia et al. 2022; Pellow 2016; Rigon and Castán Broto 2021). Recent scholarship has sought to bring these questions of inequality to the center of studies of commons and environmental projects, offering a much-needed corrective to older scholarship on environmental collective action that tended to flatten communities and adopt an overly individualizing perspective (Kashwan et al. 2021). Consideration of sociocultural history and identity is especially crucial in cities, because urban areas tend to house diverse populations and urban spaces meet diverse, parallel uses for heterogeneous populations.

45. Liboiron illustrates this through an incisive critique of a widely cited report, "Plastic Waste Inputs from Land into the Ocean," published in the journal *Science*. This research aimed to estimate the amount and source of postconsumer plastic waste entering the oceans and expressed the amount of plastic waste generated within a country in per capita figures. Liboiron points out that postconsumer waste accounts for only one area of marine plastics. They state, "Reports like these reproduce the erroneous truism that plastic pollution is a consumer problem rather than an industrial production problem" (2021, 73). The focus on consumers is accomplished through an emphasis on postconsumer or municipal waste in waste management conversations, as well as citing per capita waste measurements that erase both the role of industry in creating disposables and inequities within a region (Lepawsky 2018; Liboiron 2021).

46. Recycling is an energy-intensive industrial process that produces pollution (MacBride 2011). In the United States, where I live and teach, the myth of plastic recycling is finally being blown up by investigative journalists and incisive books like Alice Mah's *Plastic Unlimited* (Mah 2022).

47. These are two key methods of analysis employed by discard studies scholars to go beyond obvious and partial explanations, which usually tend to reinforce an existing status quo (Liboiron and Lepawsky 2022).

48. The past few years have seen a flurry of work studying questions of citizenship, inequality, race, and coloniality through and with waste infrastructures (Chalfin 2014; Doherty 2021; Fredericks 2018; O'Hare 2022; S. Sharma 2022; Stamatopoulou-Robbins 2014). I add to this body of work through my explicit focus on environmental movements and ethics in waste infrastructure.

49. At this stage, my theoretical framework was built on research in social psychology, environmental sociology, and environmental politics. In retrospect, it was rather uncritical. I was not thinking about inequality, oppression, or justice when I began this journey.

50. Over those two years, I conducted a total of fifty-five interviews with seventy individuals. Some interviews were with couples in their homes. All my interviewees self-identified as middle class.

51. I am grateful for those formative conversations with Dr. Vinay Gidwani, Dr. Solomon Benjamin, Leo Saldanha, Dr. Raka Ray, Dr. Isha Ray, Dr. Amita Baviskar, and Dr. Carol Upadhya, who pushed me to deepen and expand my inquiry. They introduced me to the scholarship on economic and urban geography, southern urbanism, and labor studies.

Chapter 1

1. The term "new middle class" is generally used in the Indian context to refer to the social groups, predominantly dominant-caste, English-speaking, and urban, who benefited from the economic liberalization policies pursued in India in the late 1980s, and whose consumption patterns and practices are globalized. In comparison, the older middle classes are thought to be the social group that benefited from state-led developmental projects such as public universities and public-sector units, though there are continuities between the older and the new middle classes. Bengaluru is home to a sizable new middle class whose members live in gated enclaves, drive cars, and indulge in discretionary spending. Further, the new middle classes are significantly more well-off than most of India's population, in that, in terms of income and consumption, they are closer to what might be described as an elite. Thus, "new middle class" operates as much as a cultural construct as a sociological term; the discourses around middle classness are as important to their self-definition as how much they earn or what they buy. For more, see Baviskar and Ray (2011), Fernandes and Heller (2006), Upadhya and Vasavi (2008). Geographer Malini Ranganathan uses the term "peripheralized middle classes" to refer to the more numerous members of Bengaluru's middle classes (compared to the smaller number of English-speaking and securely propertied new middle classes) who are "rooted in regional cultures and languages, educated, and crucially, *property owning* (though not securely so)" (Ranganathan 2014, 5). I use the terms middle class, elite middle class, and elites interchangeably, but always to refer to the well-off, urbanized, English-speaking, and culturally dominant contingent who are the subjects of this study.

2. The arguments in this chapter are based on fifty-five in-depth, semi-structured interviews with individuals who practice and promote bicycling, zero waste, and organic gardening, supplemented with participant observation and online ethnography on listservs and Facebook groups dedicated to green lifestyles in the city.

3. According to geographer Laura Pulido, ecological legitimacy attaches to a group when it is seen as a valid environmental actor, giving group members the power or moral authority to make environmental claims. Environment here is broadly understood as more-than-human life and landscape (Pulido 1996).

4. See, for example, the UK's behavioral insights team (https://www.bi.team/), partially funded by the UK Cabinet's office.

5. Consumption-based accounting of greenhouse gas emissions has been used to examine the role of individual and household consumption in climate change (Dubois et al. 2019; Mi et al. 2019; UN Environment 2020). Some of these studies demonstrate that many Global North countries and cities have reduced their territorial greenhouse gas emissions (the more traditional way of attributing greenhouse gas emissions) by simply displacing their emission-producing activities to local and global hinterlands. "Cleaning up" carbon emissions in one context is thus accomplished by discarding it elsewhere, a key technique of power as illuminated from a discard studies perspective (Liboiron and Lepawsky 2022).

6. More on this point in chapters 3 and 5, but, for a preview, see Gidwani and Reddy (2011), Schindler and Demaria (2019), M. Yates (2011).

7. The top 10 percent in turn contribute 40 percent of global greenhouse gas emissions, while the bottom half of the world's population account for only 13 percent of global greenhouse gas emissions (Chancel and Piketty 2015).

8. Emission disparities among different expenditure classes in India vary tremendously. Parikh et al. estimate that, in 2003–2004, the emissions produced by the top 10 percent of urban India (roughly 30 million people) were about fifteen times those of the bottom 10 percent of urban India, and about twenty-four times the emissions of the bottom 10 percent of rural India (Parikh et al. 2009). Using a different methodology that relied on an aggregate measure of class status combining income, occupation, and consumption profile, Michael and Vakulabharanam find that India's highest emitting class, urban elites, emit almost seven times that of the lowest emitters, rural agricultural workers (Michael and Vakulabharanam 2016). By comparison, the per capita emissions of the richest 10 per cent of US residents is almost twelve times higher than that of the richest Indians (Bhushan 2018).

9. For work on sustainable consumption as prefigurative politics, see Schlosberg and Craven (2019), L. Yates (2015).

10. Ecological citizenship, Andrew Dobson's normative theory of environmental action, posits that individuals who have historically taken more than their fair share of global resources should voluntarily compensate for this by taking on public and private actions that have beneficial outcomes for the environment (Dobson 2006).

11. This practice-based conception of class, drawn from Bourdieu's concepts of *habitus* and *field*, theorizes how class structures are reproduced by social groups through everyday practices. In Bourdieu's formulation, habitus is first shaped in the intimate context of the home, where individuals are socialized into certain ways of being and interacting with the world even as they are acquiring skills and cultural competencies. The social field (such as educational institutions or workplaces) is the setting in which these skills and dispositions are deployed and strengthened. Individuals thus build cultural capital, the combination of values, tastes, cultural goods, and

qualifications that one acquires by being in a particular social class. Cultural capital, along with economic resources and social connections, becomes the structural basis of class power and the means for creating and maintaining social distinction. Bourdieu defines the habitus as a "systems of dispositions, characteristic of the different classes and class fractions" (Bourdieu 1984, 541).

12. Carfagna et al. define eco-habitus as "a reconfiguration of high-status tastes that is part of a re-articulation of the field of high-class consumption, fostered by a more general social valorization of environmental consciousness" (Carfagna et al. 2014, 161).

13. For critical analysis of green lifestyles and sustainable consumption, see Guthman (2003), Johnston (2008), Johnston, Szabo, and Rodney (2011), Maniates (2001).

14. Bourdieu defines social capital as the "aggregate of the actual or potential resources which are linked to possession of a durable network of more or less institutionalized relationships of mutual acquaintance and recognition—or in other words, to membership in a group which provides each of its members with the backing of the collectively-owned capital" (Bourdieu 2002, 88).

15. Cultural capital, for Bourdieu, is the basis of any strategy of social distinction. He distinguishes three types of cultural capital: the embodied state (long-lasting dispositions of the mind and body), the objectified state (cultural goods), and institutionalized state (obtained through educational qualifications).

16. See also Pathak (2020).

17. Writing about the politics of slum demolition in New Delhi, Asher Ghertner argues that the "rule by aesthetics" has emerged as the primary way authorities decide what land uses are permitted or criminalized (Ghertner 2015, 6).

18. Govind Gopakumar documents a similar pattern in his study of automobility in Bengaluru. He talks about a discourse that diagnoses the rising number of automobiles as a major problem in the city. There is a contradiction here—those complain the most are often also always using automobiles; see Gopakumar (2020).

19. For a careful treatment of how greening projects became central to urban planning's efforts at creating and managing governable spaces and social improvement in multiple contexts, see Angelo (2021).

20. Gauri Pathak warns this is part of a soft-Hindutva strategy of recasting India from a secular to a Hindu-supremacist nation (Pathak 2021).

21. Gandhian nationalism in particular correlated the ability to regulate the self and public space with the ability of formal political self-rule. Garbage in this sense is a barrier to modernity, development, and progress. See Chakrabarty (1991), Hodges (2013), Kaviraj (1998), Prashad (2001).

22. For instance, just as air quality in US cities improved, it has deteriorated in China, India, and elsewhere in the Global South, as these places have become the USA's industrial heartlands, globalizing the ecological shadows of consumption (Dauvergne 2010).

23. This notion connects to Mary Douglas's framing of dirt as matter out of place. Signaling unhoused people or shelters as dirty is a means of indicating them as unacceptable within a social order, where the act of discarding maintains the order. However, as Max Liboiron and Josh Lepawsky contend, not all garbage/waste should be understood as dirt, and cleanup cannot always be conflated with purification. Sometimes cleanup is necessary to challenge power or oppression. At other moments, it becomes a tool to maintain it. Context and social relations determine when garbage functions as matter out of place (Douglas 2003; Liboiron and Lepawsky 2022).

24. As was the case in Chennai, my hometown, where communities living by the banks of the Adyar River were displaced to create an eco-park called the Adyar Poonga. For more discussion on the displacement of informal settlements and livelihoods to create green spaces, see Coelho (2020), Ellis (2011), Zimmer, Cornea, and Véron (2017).

Chapter 2

1. The name of the group has been anonymized for confidentiality.

2. The image of a dead cow elicits a particularly strong response from the crowd. Saving cows from plastics has emerged as a key concern for some fractions of dominant-caste Hindus who venerate the cow as sacred. This is likely connected to the rising expressions of Hindutva in urban environmental politics (Pathak 2021).

3. For a small sampling of the many analyses on this topic, see Agyeman et al. (2016), Barr and Devine-Wright (2012), Bulkeley and Fuller (2012), McLaren and Agyeman (2015).

4. Exemplified best by the original work of Elinor Ostrom, this work on environmental collective action explores the factors that determine whether and how communities (variously defined) successfully develop long-standing institutions (rules, codes, accountability mechanisms, and the like) needed to govern a common or shared resource in a sustainable manner (Ostrom 1990).

5. As summarized and argued in these cited studies, amongst many others: Agrawal and Gibson (1999), Blaikie (2006), DeFilippis, Fisher, and Shragge (2006), Joseph (2002).

6. In addition to care work and socially-reproductive work, the term social infrastructure is used by feminist geographers in a more expansive way to refer to all socially reproductive work that sustains life and economic enterprise (Hall 2020).

7. Aman Luthra argues that women working in the home are increasingly being conscripted into the agendas of governmental campaigns like Swachh Bharath,

which emphasizes household segregation as key to resource recovery and clean cities. This is concomitant with the increased push toward the privatization of waste processing through waste to energy plants, which demand clean feedstock of dry waste (Luthra 2020).

8. Some comparable social groups have done so in California and other parts of the United States. See MacBride (2011).

9. Gopakumar describes this agenda of taming "unruly and unsanitary" spaces with urban planning and sanitation systems while chronicling Bengaluru's landscape histories (Gopakumar 2020).

10. For a discussion of Brahmin, male dominance in urban planning, and infrastructure in the city, see Gopakumar (2020).

11. This is not unique to Bengaluru, but a phenomenon in Indian cities and beyond (Luthra 2020; Melosi 2004). This was also prevalent in Bengaluru in earlier eras, albeit at a smaller, neighborhood scale. For instance, in the 1990s, the Bengaluru Municipal Authority partnered with a civic organization called Swabhimana to recruit middle-class women into coordinating waste segregation programs in some select localities.

12. This focus on consumer action is also more broadly true of anti-plastics and zero-waste discourse more globally, which has focused on consumer plastics rather than the industrial use (and disposal) of plastics (Liboiron 2021; Mah 2022). This is despite the fact that postconsumer or municipal waste is a small part of overall waste produced in most economies, including India (Pathak 2020).

13. As part of my field research, I visited twelve apartment complexes and commercial spaces that had installed zero-waste management infrastructure such as dry waste collection and sorting centers, composting units, and biogas plants. Many of these initiatives were in relatively new gated communities that had been built in the peripheries of Bengaluru in the past decade (like in Uttarahalli, Kengeri, Jakkur, Byatranapura, Kalyan Nagar, and J. P. Nagar), while some were in older, central neighborhoods like Malleswaram, Seshadripuram, and Jayanagar. These areas generally received waste management services either directly from the BBMP or from its contractors. I also visited schemes in two newly created "revenue layouts" (unauthorized neighborhoods built on agricultural lands) that did not receive any BBMP services, and thus had self-organized for their waste management needs. I interviewed representatives from ten waste-engaged civic and nongovernmental organizations. I also monitored and engaged in social media conversations on waste management via Facebook and email listservs.

14. Here I am invoking insights from social practice theory, which has been applied widely to the study of the dynamics of everyday life, particularly in relation to questions of sustainable consumption. Practice theorists see the everyday social as embedded in and reproduced through practices that are, in turn, anchored in material

arrangements, social norms, and teleoaffective structures. Emotions are a key part of teleoaffective structures, which prominent practice theorist Theodore Schatzki describes as the pattern of orientation toward goals (teleo) and motivational engagement through emotions (affect). Emotional states associated with a practice dictate whether a particular practice is seen as right or wrong, sanctioned or transgressive; they vary by context or situatedness of the practice. Pierre Bourdieu, another theorist who put social practices as the center of his analysis of how social structures were reproduced, studied emotions as part of habitus, that is, the systems of embodied dispositions for thought and action that mediate and confront new experiences. For more on practice theory, see Reckwitz (2002), Schatzki (1996), Shove, Pantzar, and Watson (2012), and Warde (2005).

15. The term affective labor is attributed to Michael Hardt and Antonio Negri in *Empire* as a form of "immaterial labor" that they argue is increasingly channeled into productive forces in late capitalism (see Hardt and Negri 2000). However, Hardt and Negri's conceptualization has been widely challenged by feminist thinkers as lacking a serious consideration of the gendered nature of socially reproductive labor and as re-mystifying reproductive work as producing solely states of being or regimes of feeling (Federici 2008; Schultz 2006). I am deploying Hardt and Negri's term in its more narrow sense, as labor that targets the emotional states of others.

16. I provide more description of these collectives in prior publications (M. Ananth-araman 2014). More generally, "communities of practice" have been critical in changing unsustainable social practices in several contexts (Sahakian and Wilhite 2014).

17. The move away from electoral politics is often justified by people in community-based projects as a more effective and expedient way of making social change (Kennedy, Johnston, and Parkins 2018). Yet in India, middle-class place-based associations have had a long record of engaging local councilors and municipal authorities to advance their goals, and even of employing "political-society" strategies more commonly used by working-class communities (Kamath and Vijayabaskar (2014); referencing Chatterjee's (2006) distinction between working-class political society and middle-class civil society).

18. For more on this tension, see Bhan (2016), Ghertner (2011a), Roy (2009a).

19. María José Zapata Campos and Patrik Zapata note this potential of citizen-driven initiatives to infiltrate governmental schemes with radical rationales that can become activated in the future, contributing to what they call diachronic change (Zapata Campos and Zapata 2017).

Chapter 3

1. Formerly the city's central jail, Freedom Park is Bengaluru's designated home for protests and *dharnas*.

2. One of the main organizers was also the lead behind the Swachha Bangalore program that had privatized waste management in the city in the early 2000s.

3. This was quoted in a newspaper report ("Bangalore Gets Together for 'Clean-Up' Fest" 2013).

4. This "world-class" and corporate-friendly aesthetic was sometimes in tension with TCGF's messages on behavior change and citizen responsibility. I had a front seat to the debate and conflict that came up in the organizing of this event. A consulting company that worked with City Connect had hired me to document all the panel discussions. I also joined some of the event organization meetings as part of my work supporting TCGF.

5. Indeed, the only times I heard Kannada spoken at the event were when corporators made a point to speak the local language to reassert vernacular voice, or in the training sessions held for *pourakarmikas* (sanitation or conservancy workers) at the event. On the last day, a representative from the *pourakarmikas* labor union was finally invited to speak at the closing panel. When I caught up with him after the event, he was incensed. He could barely get a word in. Waste workers, whether *pourakarmikas* or waste pickers, were sidelined, silenced, patronized, and spoken down to. I wrote in my field notes, "How can a weeklong expo on waste management go with almost no involvement from the people who actually work with waste on a daily basis?" This statement reflects my then naivete of the caste and class politics of urban development in India more than anything else.

6. This is a good example of *metis* or embodied knowledge that waste workers hold. The *pourakarmikas* are right that if they have only two bins, then even if one household fails to segregate waste properly, they can no longer keep the materials separate. These categorizations of dry versus wet waste also ignore other materials like sanitary napkins or diapers that can be neither composted nor recycled. Ultimately, after much discussion and lobbying by groups representing waste workers, the rules included a category for hazardous waste.

7. Much of the high-value waste would be diverted anyway by informal workers. Estimates suggest that India recycles about 60 percent of the 5.6 million metric tons of plastic thrown away every year, compared to about 9 percent in the United States (Doron and Jeffrey 2018, 128). Waste diversion is often used as a tactic to dismiss waste picker's contributions (Luthra 2020).

8. For discussion on how racialization operates in relation to capitalism and produces particular patterns of land dispossession and environmental injustice, see Ranganathan (2016), Melamed (2015), Reddy (2021), and Pulido (2017a).

9. Srinivas (2002) does an excellent job describing Brahminical ritualistic cleaning and obsession with purity pollution. Her descriptions resonate with my own personal experiences of being reprimanded for serving myself at my grandmothers' table,

touching the pickle jar while on my period, and touching "purified" clothes (called *madi*, in Tamil) when I had not recently bathed myself.

10. Vinay Gidwani and Anand Maringanti call this work "infrastructural labor," which they argue is essential for re-creating the conditions of possibility for both urban life and capitalist enterprise (Gidwani and Maringanti 2016). Gidwani further explains that waste pickers produce what Marx called capital's general and external conditions of production through their infrastructural labor in the infra-economy (2015).

11. It is worth noting that, throughout her work, Spivak is critiquing and trying to resist the ways in which postcolonial scholarship and subaltern studies essentializes third-world subjects. Further, as Max Liboiron and Josh Lepawsky (2022, 112–113) point out, drawing on the work of Eve Tuck (2009), bell hooks (2014), and others, essentializing, when part of justice struggles, often involves using damage-centered narratives to demonstrate to those in power what type of justice or recompense is needed. Consequently, "oppressed" communities are often reduced to just that, the "oppressed." While I have certainly fallen into this trap myself in this book, my analysis is shaped by my desire not to reproduce only damage-centered narratives about waste pickers (itself an essentialized category!); for more on that see Sneha Sharma's analysis (2022, 187).

12. My illustration of waste picking, which is a broad term referring to a range of dynamic practices from sorting through mixed waste at the roadside or in transfer stations, to wading through municipal dumps and sanitary landfills to glean valuable materials, is highly abstracted. Other scholars have produced detailed and careful treatment of the general characteristics of waste-picking occupations (Dias 2016; Dias and Samson 2016; Gutberlet 2012; Porras Bulla, Rendon, and Espluga Trenc 2021; Wittmer and Parizeau 2016); its spatiality, negotiability, and social relations, which vary significantly from context to context (H. Anantharaman 2019; Butt 2019; Calleja 2021; Fahmi and Sutton 2006; Gidwani 2013a, 2015; Gidwani and Chaturvedi 2013; Gill 2007; Kornberg 2020; Samson 2010; Schenck and Blaauw 2011); and thoughtful ethnographic portrayals of the work and its communities (Millar 2018; O'Hare 2022; S. Sharma 2022).

13. Personal communication and review of data from Parisara Tanda staff who have been registering and enumerating waste pickers in the city, updated 2022. According to these staff members, Adidravidas are the largest Scheduled Caste group, many of whom are Tamil or Telugu speaking. Among Scheduled Tribes, there are Hakki-Pikki communities, some of whom specialized in specific materials like bones or hair. Published studies of the sociological composition of waste pickers are now dated. A 2010 study by Mythri Sarva Seva Samithi (MSSS) revealed that the majority of Bengaluru's waste pickers belonged to Scheduled Castes (SC), Other Backward Castes (OBC), or Scheduled Tribes (ST) including nomadic communities (CHF International and Mythri Sarva Seva Samithi 2010). In 2016, the Alliance of Indian Waste Pickers surveyed 1,896 waste pickers across the country, finding that 54 percent of the

sample identified as SC and 28 percent as ST; 68 percent identified as women. Bengaluru also has many Muslim waste pickers who are migrants from East and Northeast India (Chandran, Narayanan, and Subramanian 2019).

14. As documented in Mexico City (Guibrunet 2019), Delhi (Luthra 2019), Chennai (H. Anantharaman 2019) and more generally (Guha-Khasnobis and Kanbur 2006; Meagher 2013).

15. These modes of negotiation and practical legitimation are central to how informal work and economies persist even in the face of punitive policies or competition from more well-resourced private players (H. Anantharaman 2019; Butt 2019; Calleja 2021; Kornberg 2020; Luthra 2019).

16. Caste is also locked in through hiring practices. As Pradeep Salve and coauthors show through a study in Mumbai, municipal conservancy and sanitation work has become a hereditary occupation thanks to a policy that promotes preferential hiring for the relatives of current conservancy workers. A conservancy worker can nominate his wife, son, brother, unmarried or widowed daughter or sister, or any other dependent to the post of conservancy worker after retirement, death, or permanent disability. While this policy was created in 1972 to purportedly economically empower this community, some activists have strongly condemned it as a casteist and regressive system. The overall shortage of jobs and high levels of unemployment also push members of oppressed castes into waste occupations (Salve, Bansod, and Kadlak 2017).

17. The term resourcification offered by Hervé Corvellec and coauthors makes clear that resources are not simply already here but are the outcome of social processes that condition what is considered a resource (Corvellec et al. 2021). Indeed, waste pickers were the original pioneers of treating waste as a resource by denaturalizing it, distinguishing materials of value in discard, and, as Melanie Samson suggests, creating a new commodity frontier (Samson 2019).

18. I saw this firsthand when I attended the Rio 2012 Conference on Sustainable Development. EU think tanks and representatives like the European Environmental Agency (EEA) were particularly concerned with the issue of the world running out of vital resources and the urgent need to recover mineral resources from waste. I recall attending the premier of a documentary called *Planet Rethink*, produced by the EEA and the United Nations Environment Program, that drove home these points. In subsequent years, the European Commission has published several reports on resource scarcity and the need to transition to using waste as a resource; see European Commission (2017, 2018).

19. For more discussion on this, see Corwin (2020), Knapp (2016), Reddy (2016), Schindler and Demaria (2019).

20. The analysis is based on conflicts recorded in the Global Atlas of Environmental Injustice (Demaria and Todt 2020).

21. I first heard about the then-nascent Parisara Tanda from one of my interlocutors within the middle-class-dominated zero-waste movement in September 2012. My research at that moment was focused primarily on understanding middle-class environmental subjectivities and community-based environmental activism. I had never intended to dig into the labor question of urban environmentalism, reflecting gaps in my own education and thinking that had hidden from me the ways in which middle-class environmental activism was connected to or disconnected from the concerns of other social groups. However, my interviews with zero-waste enthusiasts and observations of community-waste management efforts had alerted me to two things: first, that ZWM systems were labor-intensive, and figuring out how to train and manage labor was a key preoccupation of zero-waste proponents (as discussed in chapter 2), and second, that ZWM members were beginning to learn about and discuss the role that the "informal sector" should or should not play in the zero-waste infrastructures they were advocating for. Soon, WIEGO hired me to work as a consultant helping the organization develop their website. I also wrote grant applications to find funding for the six staff members who at that point made up the not-yet-officially-registered organization. During that period, organizing and advocacy efforts were intense. We featured case studies of waste-picker integration in an attempt to demonstrate to Bengaluru's elites that other global cities had integrated waste pickers resulting in good environmental and public cleanliness outcomes. I played a role as an interlocutor among different community actors, writing up short case studies to share with the middle-class ZWM activists they were trying to influence. Since 2012, I have continued to support Parisara Tanda, assisting them in their internal research and external outreach activities. I have proofread reports, commented on research design, and joined the organization on some public-facing events. My positionality as someone who has supported the organization's efforts and strategies certainly influences my analysis of their work. At the same time, I aim to be critical and have posed questions to staff and leaders. This inside/outside positionality is a common challenge in community-engaged research and one that has a structuring effect on the knowledge claims produced here. Readers would do well to keep this in mind.

22. An analysis of nine informal worker movements in four countries found that the impetus for organizing informal workers largely came from the efforts of middle-class actors belonging to nongovernmental organizations (NGOs) (Chikarmane and Narayan 2005; Dias and Samson 2016). In Delhi, while the middle-class-staffed NGO Chintan trains waste pickers and represents their perspectives in policy fora (Luthra and Montieth 2021), Safai Sena, a registered group of waste pickers and small scrap dealers, has more grassroots presence and coordinates work contracts. In Pune, waste pickers are represented by a trade union called Kagad Kach Patra Kashtakari Panchayat (KKPKP) and access work opportunities through a cooperative called SWacCH, which is operated by a professional staff. More research is needed to understand how the presence of elite and middle class interlocutors in NGOs and social movements impacts modes of organizing, claims-making, and achievements, especially because recent studies have established that many movements assumed to

be mass environmental struggles, such as the famous Chipko movement, were dominated by a class of leaders (Kashwan 2022). For examples of work that examine the role of intermediary organizations and individuals, and the uneven effects of NGOs, see H. Anantharaman (2019), Dhananka (2010), Doshi (2013), McFarlane (2008), Rosaldo (2016), Schuller (2009), M. Sharma (2017a).

23. All India Central Council of Trade Unions (AICCTU) is a central trade union federation in India.

24. Starting and maintaining waste-picker organizations is a difficult project, and waste-picker organizations take many organizational forms based on local complexities. They range from trade unions, like KKPKP in Pune, to cooperatives and self-help groups to microentrepreneurs supported by NGOs. The degree to which waste pickers speak for themselves versus being represented by non-waste pickers also differs significantly (Kain et al. 2022; Rosaldo 2016). In Brazil and Colombia, waste pickers are self-organized into strong trade unions and cooperatives, while in Africa, there are more NGOs or self-help groups convening waste pickers (Kain et al. 2022; Rosaldo 2016; Samson 2016; Zapata Campos et al. 2021).

25. Parisara Tanda tried to register itself as a cooperative to follow in the footsteps of SWaCH Wastepicker Cooperative in Pune. Ultimately, however, they decided against this, because the Karnataka Souhardara Sahakari 1997 Act does not permit or support multipurpose cooperatives. When I asked staff members whether another challenge was that they had not organized enough waste pickers to function as a co-op, they said that organizing was not an issue. One member said, "The overarching climate favors companies, cooperatives are not encouraged or helped. While we are not formally a member-based organization we work like one. We are a hybrid organization that defies current classifications" (December 2018 interview at Parisara Tanda offices).

26. The organization's original mission statement reads, "Parisara Tanda, started in 2013, is an organization of waste workers that works towards improving the livelihood and quality of life of waste pickers by providing total waste management services through them to bulk generators of waste. It also provides social services and support to the waste picking community. Parisara Tanda therefore impacts both lives and the environment."

27. Twitter post, June 5, 2019, 12:06 p.m.

28. These savings were calculated based on the "avoided landfilling cost." At the time of this report, the BBMP was paying 2,219 rupees per ton of garbage dumped at the landfill to the landfill operator. The organization calculated how many tons of recyclables were being diverted by their waste pickers and used that to come up with a "savings" number.

29. Waste picking accrues value beyond what is utilized or accessible to waste pickers for their reproduction. Some of this value is accrued to economic actors higher

up the value chain. Value also accrued in the form of savings to municipality. The final form of value is the environmental value produced.

30. For a more in-depth discussion of rethinking informal work, see Millar (2018), O'Hare (2022), Tucker and Anantharaman (2020).

31. Indeed, Parisara Tanda emphasized the autonomy and independence of waste pickers as a positive attitude, and something that made them capable of puzzling out value in waste. This is akin to the narrative Katherine Millar builds out through her study of Catadores in Rio de Janiero (Millar 2018).

32. Gidwani explains the push and pull factors bringing people, especially Dalits, to cities. He argues that, in addition to economic factors, people seeking to escape oppressive social hierarchies in villages operating on the basis of caste also migrate to cities (Gidwani 2015). Millar also notes that Catadores in Brazil see the dump where they pick waste as a "refuge," providing one of the few self-employment options that enable individuals to retain some autonomy and control over their lives (Millar 2018).

33. Swetha Dhanaka Rao notes this as well, styling Bengaluru as an entrepreneurial city (Dhananka 2010).

Chapter 4

1. For a selection of studies documenting this, see Anguelovski and Connolly (2021), Bulkeley (2013), Bulkeley, Castán Broto, and Maassen (2014), Chini et al. (2017), Rosan and Pearsall (2018), Stehlin (2019). Note that some of these infrastructures are also tied to the broader project of climate adaptation and mitigation, and thus might be better described as climate urbanism.

2. For key articulations of this in the Bengaluru context, see work by Malini Ranganathan and Govind Gopakumar (Ranganathan 2014; Gopakumar 2020).

3. De-risking describes the financial and policy processes through which private investors as well as government authorities attempt to decrease the downside risks of investments. Strategies can include passing off costs onto third parties, changing policies to reduce barriers to technology deployment, or floating new spin-off companies or public-private partnerships to protect the main agency or company from bankruptcy (Schmidt 2014).

4. See, for example, recent work on the link between bicycling infrastructure, urban greening, and gentrification (Checker 2011; Gould and Lewis 2016; Hoffmann and Lugo 2014), and also broader, more systemic critiques (Castán Broto and Westman 2019; Luque-Ayala, Marvin, and Bulkeley 2018; Ramaswami 2020).

5. Slum settlements were forcibly removed under the guise of flood-prevention efforts in Mumbai and public transit networks in Chennai (Doshi 2019; Narayan 2015).

6. Green here refers to "nature in the city . . . in its more verdant, simplistic and realistic form," while gray refers to "the concept of social, technological urban space as inherently sustainable," or the promise of technical expertise, efficiency, and ultimately, modernity (Wachsmuth and Angelo 2018, 1040).

7. Key examples of studies of infrastructuring that have influenced my own approach to studying these everyday processes include Gopakumar (2020), Millington and Scheba (2020), Stehlin (2019).

8. A situated urban political ecology approach, articulated initially by Lawhon, Ernstson, and Silver (2014), differs from traditional "Marxist" urban political ecology in that it emphasizes exploring how notions of people as infrastructure, embodied experience, and situated knowledge are central to urban political ecologies, as opposed to starting with an abstract Marxist notion of power. It also considers the ways in which socially reproductive work (discussed in chapter 2) functions as social infrastructure that enables capitalist enterprise as well as life in the city.

9. The concept of heterogeneous infrastructural configurations suggests a rethinking of infrastructure from cities of the Global South, enabling a "clearer analysis of infrastructural artefacts not as individual objects but as parts of geographically spread socio-technological configurations: configurations which involve many different technologies, relations, capacities and operations, entailing different risks and power relationships" (Lawhon et al. 2018, 723).

10. See Doherty (2021), Fredericks (2018), Graham and McFarlane (2014), Silver (2014), Stamatopoulou-Robbins (2019).

11. For a selection of infrastructural citizenship studies, see Anand (2017), Fredericks (2018), Moore (2012), Ranganathan (2014).

12. There are no agreed-upon numbers regarding the number of operating DWCCs. The BBMP claims to have commissioned 189 according to its SWM data portal as of 2022.

13. I traced the DWCC in several ways. I visited some of the earliest iterations during fieldwork in 2012–2013. In 2018, I visited an upgraded DWCC operated by waste pickers affiliated with Parisara Tanda. From 2012 to 2019, I created a database of newspaper articles and social media posts about DWCC. I conducted document content analysis on this database to trace both how the material forms and the perception of the DWCC shifted. I also rely on waste-picker- and Parisara-Tanda-produced analyses and reports, alongside information from the BBMP website.

14. For an analysis of the conflicts and politics around incineration in Delhi, see Demaria and Schindler (2016), Kornberg (2019b).

15. The Clean Development Mechanism was a financing tool designed to attract financing for carbon mitigation projects in the developing world while simultaneously

enabling top polluters to meet climate targets without changing domestic industrial or consumption conditions. It is among the many carbon finance tools that apply market principles to explore achieving climate mitigation at least cost (Newell and Paterson 2010). Like many other schemes, its local environmental impacts are considered secondary to overall aggregate outcomes of greenhouse gas reduction (Hesketh 2022).

16. Assa Doron and Robin Jeffrey offer a careful discussion of the challenges of implementing incineration technologies in India in their treatise on waste in India (Doron and Jeffrey 2018, 146–153).

17. Middle-class organizations frequently allude to state/bureaucratic corruption as a reason that Indian cities continue to be plagued by infrastructural and environmental issues. These simplistic narratives are used to sometimes push for greater privatization of infrastructures. A critical analysis of these corruption narratives and their intersections with classed and raced privilege is presented by Sapana Doshi and Malini Ranganathan (Doshi and Ranganathan 2017, 2019).

18. This quote is from a slide deck presented by a ZWM member in September 2012.

19. Though municipal recycling is in many cases just the displacement of burdens of waste processing to faraway places (Gregson et al. 2016; Heiges and O'Neill 2022; MacBride 2011).

20. Decentralization, self-reliance, and localism are common tropes in middle-class civic and sustainability politics, both in India and beyond, a framing that the state propagates via its governmental schemes (Anjaria 2009; Hébert and Mincyte 2014; Roy 2009a). See chapter 3 for more.

21. Numbers have been calculated by the author based on newspaper articles and communication (BBMP 25 lakhs/center * 200 centers = 50 crore rupees. Three waste-to-energy plants = 600 crore rupees).

22. This explanation for the benefits of coproduction of urban services is aligned with the perspective offered by Elinor Ostrom and her followers, who argue that coproduction can have clear economic benefits (more efficient, less costly), while also reinvigorating greater participation in civic life through communitarian engagement (Moretto et al. 2018; Ostrom 1993, 1996). Durose and Richardson further contend that coproductive approaches might be more effective at solving "wicked problems" (such as garbage), where public policy-led approaches have had limited success (Durose and Richardson 2015).

23. ITC runs a corporate social responsibility scheme then called Wealth out of Waste (WoW), which operates across several Indian cities. Motivated by a legislation that requires all companies to divert 2 percent of their profits to CSR schemes, ITC also saw a win-win in obtaining access to paper for its paper mills (for more information about this scheme, see Doron and Jeffrey 2018, 203–205). In Bengaluru, they

collected dry recyclable waste from large apartment complexes every week, paying households the nominal sum of 2 rupees/kg of mixed dry waste. The company had invested in equipment and had tie-ins with local waste collection contractors and NGOs for labor. In 2017, ITC, in partnership with a diverse collection of NGOs, operated somewhere around sixty dry waste collection centers in Bengaluru, for example, in partnership with the Samarthanam Trust for the Blind.

24. In 2022, thirty-nine DWCCs were operated by waste pickers and scrap dealers. These numbers have fluctuated. The highest number was forty-four just before the pandemic (review of data from Parisara Tanda). In 2022, I met a waste picker who had left her DWCC just as the pandemic hit. She explained that she realized that, with everything shut down, the price of plastics was going to fall with the price of oil. So, she decided to cut her losses and go back to her village in Tamil Nadu for a while. She was now back in Bengaluru and about to start working as a sorter in a new waste processing center commissioned by Parisara Tanda and partners.

25. According to the BBMP dashboard, in 2021, approximately 58.5 percent of the city's waste generation is from households, 49.7 percent from commercial establishments, and 6.8 percent from street-sweeping. The per capita waste generation from regular households in the city is 309g/day, and that from slums is 300g/day. (See https://apps.bbmpgov.in/swmreports/. Last accessed 10/11/2022.)

26. Though noting that DWCCs are not required to collect or process waste from bulk generators like hotels, hospitals, or apartment complexes, which, according to the 2012 guidelines, are supposed to have in-house storage and sorting systems. Yet, DWCC operators are often asked to service these spaces too.

27. Liboiron and Lepawsky introduce the term scalar mismatch to discuss how partial knowledge undermines effective, appropriate, and just responses to dysfunction in waste systems. Scalar mismatch, in their words, is "where one instance is taken to be the whole phenomenon or where one perspective is assumed to work in all cases" (Liboiron and Lepawsky 2022, 39). This broader point about the situated nature of knowledge, and the concomitant need to consider how positionality affects research and knowledge claims, is a central contribution of feminist standpoint theory and Black feminist thought (P. H. Collins 2002; Harding 2004). For a connection to critical sustainability and environmental studies, see Stephens (2020) and Castán Broto and Westman (2019, 58–63).

28. The public interest litigation demanded that DWCCs not be added into the general tenders for waste collection and street-sweeping.

29. For a testimony about the challenges posed by COVID-19 from a waste picker who operates a DWCC in Bengaluru see https://bengaluru.citizenmatters.in/covid -19-lockdown-waste-collectors-masks-gloves-segregation-hazard-dwcc-indira-canteen -44966.

30. The radio-show interviews are conducted in Tamil or Kannada. I listened to the radio shows, transcribed them, and translated these excerpts from Tamil/Kannada into English. They are shared with the permission of Radio Active.

31. This decision was codified in 2020 in the BBMP By-Laws on Solid Waste Management.

32. This decision was driven by the Karnataka state government, which in contravention of several acts and laws decided to set up a new corporation called the Bengaluru Solid Waste Management Company to handle waste management in the city (Menezes 2022a). This move is undemocratic, as it moves waste management systems further away from the control of local governments and citizen committees to instead increase the power of a small number of technocrats. As its first move, this company proposed scrapping the existing system in which dry and wet wastes are collected separately, and instead moved to appoint a single agency to collect all types of waste. This model emulates Indore, celebrated by the central government as India's cleanest city for four consecutive years. Yet, waste pickers lost out in the Indore model (global_rec 2018).

33. A 2021 study found that only 15 percent of the city's dry waste is processed in these centers, with a majority still being dumped in peri-urban areas (Sensing Local 2022). With the closure of Mandur and Mavallipura, new dumps were set up in abandoned quarries and in *gommala* lands (Akshatha 2019). Bengaluru has also commissioned waste processing centers in villages outside the city, much to the chagrin of local residents, who complain about smell and leachates from unsegregated waste dumped outside some of these centers (Bhat 2020; Doron and Jeffrey 2018, 194–198; Prasher 2022).

Chapter 5

1. Nonresident Indians (NRIs) have a large presence in cities like Bengaluru and Hyderabad. The two cities host offices of several multinational IT corporations like Microsoft, Google, and Wipro. Anna Lee Saxenian has brought our attention to the role of NRIs as facilitators of trade with and investment in their countries of origin. Moreover, in the 2010s, there was a trend in NRIs moving back home. They bring with them connections to investors abroad as well as new social and environmental values. Ex-NRIs are also well-represented in the eco-lifestyle communities I studied (M. Anantharaman 2017; E. Chacko 2007; Saxenian 2002, 2005).

2. I find the use of "virgin" to refer to materials in this manner very sexist and antiquated. However, this is the dominant term used in these spaces.

3. For some examples of policy and journal articles pointing to the win-win promises of the circular economy, see Ellen MacArthur Foundation (2016), European Commission (2018), Gower and Schröder (2016), McGinty (2020), Stahel (2016).

For example, the World Bank in its "What a Waste 2.0" report argues that, when waste pickers are properly supported and organized, informal recycling can create employment, improve local industrial competitiveness, reduce poverty, and decrease municipal spending (Kaza et al. 2018). Melanie Samson documents this trend from her vantage point as a key scholar-practitioner affiliated with WIEGO (Women in Informal Employment: Globalizing and Organizing) (Samson 2019). For in-depth, critical review of circular economy discourse and practice, see M. Anantharaman (2021), Corvellec, Stowell, and Johansson (2022).

4. My analytic here sees the local and global as a continuum. This story has always implicated and invoked ideas, people, and places beyond Bengaluru. In this chapter, I show how what is happening in Bengaluru is in turn creating the conditions of possibility for imagined and hoped-for global circular economy transitions. The local is not simply affected by global forces. This, for example, has been posited in the "impact model of development" (Parnell and Robinson 2012), but here I understand the local-global as co-constitutive sites of friction (Tsing 2011).

5. At the 2020 EU-India Summit on Cooperation, the two parties issued a declaration in support of "enhancing resource efficiency and moving towards a more circular economic model that reduces primary resource consumption, striving towards non-toxic material cycles, and enhances the use of secondary raw materials," as part of a cooperative act to achieve Paris Climate targets and sustainable development goals (Barczak 2022).

6. It is important to note that waste and scrap trade is multidirectional, and, globally, can involve South-South trade as well as trade from South to North (Lepawsky 2018).

7. A similar rationale was offered for the Bo2W approach that UN StEP (Solving the E-Waste Problem) attempted to institute. The Bo2W approach tried to institute a commodity chain between high-income and low-income countries to leverage low-cost labor for manual disassembly in low-income countries and high-technology refineries in the former. According to Lepawsky et al., "Such a commodity chain is, according to Bo2W, a 'win-win' scenario since workers at manual disassembly plants would accrue benefits such as higher wages and improved occupational health and safety while high-technology refining facilities would gain access to low-cost but high-quality feedstock that is superior in purity to that derived from automated shredding of discarded electronics" (Lepawsky et al. 2017).

8. Several case studies of integration or inclusion have been analyzed in the literature, with the full range of emotions—from celebratory accounts to ones deeply critical of the whole enterprise to nuanced analysis in between (Gower and Schröder 2016; Gutberlet 2012; Gutberlet and Carenzo 2020; O'Hare 2020; Reddy 2015; Rosaldo 2019; Samson 2019).

9. I take a lot of inspiration here from the fantastic work of Melanie Samson (2015).

10. David Harvey uses the phrase accumulation by dispossession to refer to the continuation and proliferation of so-called primitive accumulation processes in the neoliberal era through privatization, financialization, and the management and manipulation of crisis, as well as through changes to the redistributive mechanisms of the state (Harvey 2010).

11. For a selection of studies on waste-picker activism strategies and wins, see Chintan (2012), Dias (2016), Gutberlet (2021), Kain et al. (2022), Rosaldo and Alegre (2016).

12. For a thorough discussion of the diverse forms of integration efforts, from those seeking to bring waste pickers into waged relationships to those that emphasize contracting with informal recyclers, see Samson (2019). Several other cases of integration have now been studied in the literature, covering diverse cities in Asia, Africa, and Latin America (Fergutz, Dias, and Mitlin 2011; Gutberlet et al. 2016; Manzi, Santana, and Marchi 2022; O'Hare 2020; Rosaldo 2019).

13. Primitive accumulation "freed" agricultural or other self-provisioning populations as proletarians, i.e., waged laborers who then have the surplus value they generate through their labor extracted as profit (Glassman 2016).

14. As EPR 2020 guidelines state, "Secondly, an important factor which is indirectly contributing to the cleanliness of the city are the rag pickers/assemblers/recyclers. They are anyway contributing to the mechanism of EPR without any benefit. This fraction of the stakeholders should be supported for the better management of the waste under the mechanism of EPR." See http://moef.gov.in/wp-content/uploads /2020/06/Final-Uniform-Framework-on-EPR-June2020-for-comments.pdf.

15. For a detailed description of the chains of resource recovery and value addition that span formal and informal sectors, see Doron and Jeffrey (2018, 211–231), Gill (2009).

16. See Thompson (2019).

17. The Body Shop, a multinational brand, has long styled itself as more ethical and eco-friendly than its competitors. In 2018, it committed to increasing the use of recycled plastic in its products (Purkayastha and Fernando 2007). On the company's website, we find a spirited invocation to corporate responsibility: "We want to find a use for the plastic that already exists in the world. By using Community Fair Trade recycled plastic from India, where almost a third of waste goes uncollected, we're fighting for people and the planet by supporting plastic 'waste pickers' with access to more sanitary working conditions, a fair price and the respect and recognition they deserve."

18. See Slavin (2019).

19. As described on their website (https://www.plasticsforchange.org/offset).

20. As described on their website (https://foundation.plasticsforchange.org/).

21. The NGOization of waste-picker struggles has produced conflict and debate, as recapped by Manuel Rosaldo (Rosaldo 2016).

22. It is worth reminding the reader here that the climate for claims-making is very hostile in Bengaluru today, particularly for Muslims and Dalits. In 2019, when the Citizenship Amendment Act was passed by the national government, de facto threatening to deprive many Muslims of their citizenship rights, some migrant waste-picking communities came under scrutiny and pressure. Migrant waste pickers from West Bengal (a state in India) were accused of being Bangladeshis and kicked out of their homes (AICCTU 2021).

23. In a 2013 interview, one member of Parisara Tanda's staff said the following:

> But we are looking to build competency within waste pickers themselves because we know that it's a very important source of their livelihood. However, obviously we cannot employ all of them because dry waste collection is 200 of them. Even if you take five employees, it is just a thousand waste pickers that we have employed. So, we are trying to get these arrangements wherever we can actually. In apartments when you go and they said "oh we need housekeeping staff for segregation," we say, "why not employ the waste pickers that are actually supposed to come here and then segregate and then they'll get the money for whatever they do?" So some apartments have been open to that. We are trying to put them into wherever we feel that they can. (Interview with Parisara Tanda staff member, 2013)

24. See MindTree (n.d.), Mindtree Ltd. (2014).

25. Parisara Tanda invited loan officers from banks to visit DWCCs to impress upon them the economic viability of waste-picker-run businesses. Middle-class staff cosign loans with waste pickers. Using these strategies, they were successful in getting one lender to lower their interest rates from a prohibitive 15 percent to a more manageable 8 percent. As one Parisara Tanda activist told me in 2018, "The rich don't have to pay back their loans and will get unlimited lines of credit. The poor are always scrutinized. But our loan repayment rate is 98 percent. Right now, waste pickers have to spend 10 percent of their earnings on paying back loans they took. This needs to change."

26. Personal communication from staff members, 2020.

27. Personal communication and estimate by Parisara Tanda staff, 2022.

28. Melanie Samson terms this "epistemic dispossession" and "epistemic injustice." The epistemic injustice lies in not recognizing informal workers as producers of knowledge. Dispossession occurs when the knowledge is appropriated by private enterprises without compensation or even permission (Samson 2015).

29. I am invoking Raj Patel and Jason Moore's concept of cheapening, which they argue is a core mechanism through which capitalism overcomes its internal and external crises. "Cheap is a strategy, a practice, a violence that mobilizes all kinds of work—human and animal, botanical and geological—with as little compensation as possible" (Patel and Moore 2018, 22).

30. This vignette draws on the ethnographic fieldnotes I compiled while attending the 4th UNEA Assembly in 2019 as a panelist on a side-event. This specific panel that I recount was open to civil society observers such as myself.

31. Doherty here is referencing the findings from a widely cited study (Jambeck et al. 2015). Max Liboiron in *Pollution Is Colonialism* calls into question the quality and coverage of the data used in the study, something that the report authors also acknowledge. A major issue in this study was that it failed to consider import and export of waste and flattens both global flows and intranational differences by citing nation-level statistics. To quote Liboiron, "It is simply impossible math, pure charisma" (2021, 74). This study has been used by countries like the United States to shirk responsibility for plastic pollution, as the UNEA experience demonstrates.

32. "Plastic can be handled if it is segregated and recycled properly," said Vijay Kumar V, President, Karnataka State Plastic Association (KSPA), in a 2020 interview with the Economic Times (Shekhar 2018). Gauri Pathak and coauthor note how the plastic industry in India took out ads to delay bans (Pathak and Nichter 2021).

33. Investigative journalist Amy Westervelt produced a four-part series on her podcast *Drilled* documenting connections between plastic use and the fracking boom. In an October 21 tweet, she summarizes the findings of the series: "TLDR; demand for fossil fuels in the transport and residential sectors are declining. Oil cos need to make that profit up somewhere and they've chosen to do it in petrochemicals . . . so, plastic. It's a two-fer because they can make it with excess fracking crap."

34. For the full text of the declaration for the meaningful inclusion and recognition of waste pickers and other informal actors, see the Global Alliance of Wastepickers (global_rec 2022).

35. This is thoughtfully demonstrated in several studies that adopt a critical perspective on waste and systems of wasting (Arefin 2019; Doherty 2021; Liboiron 2021; Pellow 2004; Stamatopoulou-Robbins 2019).

Conclusion

1. Deepak is referring to a policy decision by Narendra Modi's BJP government that overnight rendered most of the currency circulating in the economy worthless. While the policy was purportedly designed to flush out "black money," it had devastating impacts on the cash-oriented informal and rural economies of India.

2. I was for a long time in my life growing up in a Brahmin, middle-class household and community. I had naturalized these divisions and hierarchies in my life.

3. In October 2021, the BBMP Pourakarmika Sangha organized a citywide rally to protest unpaid and inadequate wages; disrespectful treatment by BBMP managers, contractors, and Bengaluru residents; and inhumane working conditions in what is essentially a caste- and gender-based urban occupation (*The Hindu* 2021b).

4. In the closing chapter of their key text on discard studies, Max Liboiron and Josh Lepawksy assert changing discarding means posing the question, "How can we discard well? And what does 'well' mean, and to whom?" Their question proceeds from their analysis that all systems must discard, and that there is no real "away" for waste (Liboiron and Lepawsky 2022, 125–133).

5. I was that child. Between the ages of six and when I left India at age twenty, I contracted malaria six times. Each time I became emaciated. The mosquitoes that grew aplenty in the waste-choked Cooum River next to my house loved me. My mother was asthmatic and allergic to every mosquito-killing chemical out there. My unfortunate nickname in school was Malaria. This formative experience was my first lesson in political ecology—there are no privatized, consumer-centric solutions to shared environmental problems.

References

Ablett, J., A. Baijal, E. Beinhocker, A. Bose, D. Farell, U. Gersch, E. Greenber, S. Gupta, and S. Gupta. 2007. "The 'Bird of Gold': The Rise of India's Consumer Market." McKinsey Global Institute report. May 1, 2007. https://www.mckinsey.com/featured -insights/asia-pacific/the-bird-of-gold.

Agarwala, Rina. 2013. *Informal Labor, Formal Politics, and Dignified Discontent in India.* Cambridge Studies in Contentious Politics. Cambridge, UK: Cambridge University Press. https://doi.org/10.1017/CBO9781139198738.

Agrawal, Arun. 2005. "Environmentality: Community, Intimate Government, and the Making of Environmental Subjects in Kumaon, India." *Current Anthropology* 46 (2): 161–190. https://doi.org/10.1086/427122.

Agrawal, Arun, and Clark C. Gibson. 1999. "Enchantment and Disenchantment: The Role of Community in Natural Resource Conservation." *World Development* 27 (4): 629–649. https://doi.org/10.1016/S0305-750X(98)00161-2.

Agyeman, Julian. 2013. *Introducing Just Sustainabilities: Policy, Planning, and Practice.* London: Zed Books.

Agyeman, Julian, Robert D. Bullard, and Bob Evans. 2002. "Exploring the Nexus: Bringing Together Sustainability, Environmental Justice and Equity." *Space and Polity* 6 (1): 77–90. https://doi.org/10.1080/13562570220137907.

Agyeman, Julian, and Tom Evans. 2003. "Toward Just Sustainability in Urban Communities: Building Equity Rights with Sustainable Solutions." *ANNALS of the American Academy of Political and Social Science* 590 (1): 35–53. https://doi.org/10.1177 /0002716203256565.

Agyeman, Julian, David Schlosberg, Luke Craven, and Caitlin Matthews. 2016. "Trends and Directions in Environmental Justice: From Inequity to Everyday Life, Community, and Just Sustainabilities." *Annual Review of Environment and Resources* 41:321–340. https://doi.org/10.1146/annurev-environ-110615-090052.

AICCTU. 2021. "Migrant Waste Pickers of Bengaluru—Social Exclusion and Deprivation of Dignity of Labor." AICCTU. http://aicctu.org/workers-resistance/v1/workers

-resistance-august-2021/migrant-waste-pickers-bengaluru-%E2%80%93-social
-exclusion-and-deprivation-dignity-labor.

Aiyadurai, Ambika, and Prashant Ingole. 2021. "Invisibility of Caste in Environmental Studies." *Indian Express*, November 29, 2021. https://indianexpress.com/article
/opinion/columns/invisibility-of-caste-in-environmental-studies-7644987.

Akshatha, M. 2019. "Bengaluru Dumping Its Garbage in Nearby Villages." *Economic
Times*, August 9, 2019. https://economictimes.indiatimes.com/news/politics-and-nation
/bengaluru-dumping-its-garbage-in-nearby-villages/articleshow/70599664.cms
?from=mdr.

Alkon, Alison Hope. 2012. *Black, White, and Green: Farmers Markets, Race, and the Green
Economy*. Geographies of Justice and Social Transformation. Athens: University of
Georgia Press.

Ambedkar, B. R. 2014. *Annihilation of Caste: The Annotated Critical Edition*. London:
Verso Books.

Amorim-Maia, Ana T., Isabelle Anguelovski, Eric Chu, and James Connolly. 2022.
"Intersectional Climate Justice: A Conceptual Pathway for Bridging Adaptation
Planning, Transformative Action, and Social Equity." *Urban Climate* 41 (January):
101053. https://doi.org/10.1016/j.uclim.2021.101053.

Anand, Nikhil. 2015. "Leaky States: Water Audits, Ignorance, and the Politics of
Infrastructure." *Public Culture* 27 (2): 305–330. https://doi.org/10.1215/08992363
-2841880.

Anand, Nikhil. 2017. *Hydraulic City: Water and the Infrastructures of Citizenship in
Mumbai*. Durham, NC: Duke University Press.

Anand, Nikhil, Akhil Gupta, and Hannah Appel. 2018. *The Promise of Infrastructure*.
Durham, NC: Duke University Press.

Anantharaman, Harsha. 2019. "Neo-Liberalising Inclusion?" *Economic and Political
Weekly* 54 (47): 61. https://www.epw.in/journal/2019/47/review-urban-affairs/neo
-liberalising-inclusion.html.

Anantharaman, Manisha. 2014. "Networked Ecological Citizenship, the New Middle
Classes and the Provisioning of Sustainable Waste Management in Bangalore, India."
Journal of Cleaner Production (Special Volume: Sustainable Production, Consumption
and Livelihoods: Global and Regional Research Perspectives) 63 (January): 173–183.
https://doi.org/10.1016/j.jclepro.2013.08.041.

Anantharaman, Manisha. 2017. "Elite and Ethical: The Defensive Distinctions of
Middle-Class Bicycling in Bangalore, India." *Journal of Consumer Culture* 17 (3): 864–886.
https://doi.org/10.1177/1469540516634412.

Anantharaman, Manisha. 2021. "Reclaiming the Circular Economy." In *The Oxford
Handbook of Comparative Environmental Politics*, edited by Jeannie Sowers, Stacy Van

Deveer, Erika Weinthal. Oxford: Oxford University Press. https://doi.org/10.1093/oxfordhb/9780197515037.013.30.

Anantharaman, Manisha, Emily Huddart Kennedy, Lucie Middlemiss, and Sarah Bradbury. 2019. "Who Participates in Community-Based Sustainable Consumption Projects and Why Does It Matter? A Constructively Critical Approach." In *Power and Politics in Sustainable Consumption Research and Practice*, edited by Cindy Isenhour, Mari Martiskainen, Lucie Middlemiss, 178–200. Abingdon, UK: Routledge.

Anantharaman, Manisha, and Patrick Schroeder. 2021. "Opinion: Why We Need to Fund the Circular Economy." Devex. September 13, 2021. https://www.devex.com/news/sponsored/opinion-why-we-need-to-fund-the-circular-economy-101554.

Angelo, Hillary. 2021. *How Green Became Good: Urbanized Nature and the Making of Cities and Citizens*. Chicago: University of Chicago Press.

Anguelovski, Isabelle. 2015. "Alternative Food Provision Conflicts in Cities: Contesting Food Privilege, Injustice, and Whiteness in Jamaica Plain, Boston." *Geoforum* 58 (January): 184–194. https://doi.org/10.1016/j.geoforum.2014.10.014.

Anguelovski, Isabelle, and James J. T. Connolly. 2021. *The Green City and Social Injustice: 21 Tales from North America and Europe*. London: Routledge.

Anjaria, Jonathan Shapiro. 2009. "Guardians of the Bourgeois City: Citizenship, Public Space, and Middle-Class Activism in Mumbai 1." *City and Community* 8 (4): 391–406. https://doi.org/10.1111/j.1540-6040.2009.01299.x.

Annepu, R. K. 2012. "Report on Sustainable Solid Waste Management in India." *Waste-to-Energy Research and Technology Council (WTERT)*, 1–189. Master's Thesis. https://gwcouncil.org/m-s-thesis-sustainable-solid-waste-management-in-india.

Appadurai, Arjun. 2001. "Deep Democracy: Urban Governmentality and the Horizon of Politics." *Environment and Urbanization* 13 (2): 23–43. https://doi.org/10.1177/095624780101300203.

Arefin, Mohammed Rafi. 2019. "Infrastructural Discontent in the Sanitary City: Waste, Revolt, and Repression in Cairo." *Antipode* 51 (4): 1057–1078. https://doi.org/10.1111/anti.12562.

Ball, Sam, dir. 2021. *Garbage Piles and Violent Clashes: The Waste Crisis Roiling Tunisia's Sfax*. Video. France 24. https://www.france24.com/en/video/20211110-garbage-piles-and-violent-clashes-the-waste-crisis-roiling-tunisia-s-sfax.

"Bangalore Gets Together for 'Clean-Up' Fest." 2013. *Citizen Matters, Bengaluru* (blog). February 3, 2013. https://bengaluru.citizenmatters.in/4879-bangalore-looks-at-optimal-waste-management-solutions-4879.

Banks, Nicola, Melanie Lombard, and Diana Mitlin. 2020. "Urban Informality as a Site of Critical Analysis." *Journal of Development Studies* 56 (2): 223–238. https://doi.org/10.1080/00220388.2019.1577384.

Barczak, Piotr. 2022. "Circular Economy: A Win-Win for EU-India Cooperation and Trade." *META* (blog). May 19, 2022. https://meta.eeb.org/2022/05/19/circular-economy-a-win-win-for-eu-india-cooperation-and-trade.

Barendregt, Bart, and Rivke Jaffe. 2014. *Green Consumption: The Global Rise of Eco-Chic.* London: Bloomsbury.

Barkin, David. 2022. "Shaping a Communitarian Ethos in an Era of Ecological Crisis." *Frontiers in Sustainability* 3. https://doi.org/10.3389/frsus.2022.944252.

Barnett, Erica C. 2019. "Bathtubs, Paint, and Couches Land in Seattle's Homeless Encampments. Guess Who Gets Punished?" *Grist*, July 9, 2019. https://grist.org/article/seattle-homeless-encampments-growing-trash-problem.

Barr, Stewart, and Patrick Devine-Wright. 2012. "Resilient Communities: Sustainabilities in Transition." *Local Environment* 17 (5): 525–532. https://doi.org/10.1080/13549839.2012.676637.

Baviskar, Amita. 2005. "Red in Tooth and Claw? Looking for Class in Struggles over Nature." In *Social Movements in India: Poverty, Power and Politics*, edited by Raka Ray and Mary Fainsod Katzenstein, 161–178. Lanham, MD: Rowman and Littlefield.

Baviskar, Amita. 2011. "Cows, Cars and Cycle-Rickshaws: Bourgeois Environmentalism and the Battle for Delhi's Streets." In *Elite and Everyman: The Cultural Politics of the Indian Middle Classes*, edited by Amita Baviskar and Raka Ray, 391–418. New Delhi: Routledge.

Baviskar, Amita. 2019a. "City Limits: Looking for Environment and Justice in the Urban Context." In *Rethinking Environmentalism: Linking Justice, Sustainability, and Diversity,* edited by Sharachchandra Lele, Eduardo S. Brondizio, John Byrne, Georgina M. Mace, Joan Martinez-Alier, 85–97. Cambridge, MA: MIT Press.

Baviskar, Amita. 2019b. *Uncivil City: Ecology, Equity and the Commons in Delhi.* New Delhi: Sage India.

Baviskar, Amita, and Raka Ray, eds. 2011. *Elite and Everyman: The Cultural Politics of the Indian Middle Classes.* New Delhi: Routledge.

Bayat, Asef. 2000. "From 'Dangerous Classes' to 'Quiet Rebels': Politics of the Urban Subaltern in the Global South." *International Sociology* 15 (3): 533–557. https://doi.org/10.1177/026858000015003005.

BBC News. 2018. "Twenty Million Indians Apply for 100,000 Railway Jobs." March 27, 2018, sec. India. https://www.bbc.com/news/world-asia-india-43551719.

Beary, Habib. 2012. "Bangalore Stinks as Garbage Piles Up in Indian City." *BBC News*, August 29, 2012. http://www.bbc.com/news/world-asia-india-19407115.

Bhalla, Nita, and John Ndiso. 2019. "U.S. Weakens First Global Commitment on Curbing Single-Use Plastics." *Reuters*, March 15, 2019, sec. Emerging Markets. https://www.reuters.com/article/us-global-plastics-pollution-idUSKCN1QW2J7.

Bhan, Gautam. 2009. "'This Is No Longer the City I Once Knew': Evictions, the Urban Poor and the Right to the City in Millennial Delhi." *Environment and Urbanization* 21 (1): 127–142. https://doi.org/10.1177/0956247809103009.

Bhan, Gautam. 2016. *In the Public's Interest: Evictions, Citizenship, and Inequality in Contemporary Delhi*. Athens: University of Georgia Press.

Bhat, Mrinalini. 2020. "Waste Plant Raises a Stink near Electronics City." *Times of India*, November 3, 2020. https://timesofindia.indiatimes.com/city/bengaluru/waste -plant-raises-a-stink-near-electronics-city/articleshow/79011074.cms.

Bhattacharya, Tithi. 2017. *Social Reproduction Theory: Remapping Class, Recentering Oppression*. Mapping Social Reproduction Theory. London: Pluto Press.

Bhushan, Chandra. 2018. "Rich Indians versus Rich (and Poor) Americans." Commentary. Center for Science and Environment. https://www.cseindia.org/a-commentary -on-consumption-rich-indians-versus-rich-and-poor-americans-9019.

Bigger, Patrick, Jessica Dempsey, Adeniyi P. Asiyanbi, Kelly Kay, Rebecca Lave, Becky Mansfield, Tracey Osborne, Morgan Robertson, and Gregory L. Simon. 2018. "Reflecting on Neoliberal Natures: An Exchange." *Environment and Planning E: Nature and Space* 1 (1/2): 25–75. https://doi.org/10.1177/2514848618776864.

Blaikie, Piers. 2006. "Is Small Really Beautiful? Community-Based Natural Resource Management in Malawi and Botswana." *World Development* 34 (11): 1942–1957. https://doi.org/10.1016/j.worlddev.2005.11.023.

Bonatti, Valeria, and Zsuzsa Gille. 2019. "Changing Registers of Visibility: Immigrant Labor and Waste Work in Naples, Italy." *International Labor and Working-Class History* 95:114–129. https://doi.org/10.1017/S0147547919000085.

Bourdieu, Pierre. 1984. *Distinction: A Social Critique of the Judgement of Taste*. Cambridge, MA: Harvard University Press.

Bourdieu, Pierre. 2002. "The Forms of Capital." In *Readings in Economic Sociology*, edited by Nicole Woolsey Biggart, 280–291. Hoboken, NJ: Blackwell. http://onlinelibrary.wiley .com/doi/10.1002/9780470755679.ch15/summary.

Braun, Yvonne A., and Assitan Sylla Traore. 2015. "Plastic Bags, Pollution, and Identity: Women and the Gendering of Globalization and Environmental Responsibility in Mali." *Gender and Society* 29 (6): 863–887. https://doi.org/10.1177/0891243215602101.

Breman, Jan, and Marcel van der Linden. 2014. "Informalizing the Economy: The Return of the Social Question at a Global Level." *Development and Change* 45 (5): 920–940. https://doi.org/10.1111/dech.12115.

Bulkeley, Harriet. 2013. *Cities and Climate Change*. Abingdon, UK: Routledge.

Bulkeley, Harriet, Vanesa Castán Broto, Mike Hodson, and Simon Marvin. 2010. *Cities and Low Carbon Transitions*. Abingdon, UK: Routledge.

Bulkeley, Harriet, and Vanesa Castán Broto. 2014. "Urban Experiments and Climate Change: Securing Zero Carbon Development in Bangalore." *Contemporary Social Science* 9 (4): 393–414. https://doi.org/10.1080/21582041.2012.692483.

Bulkeley, Harriet, Vanesa Castán Broto, and Anne Maassen. 2014. "Low-Carbon Transitions and the Reconfiguration of Urban Infrastructure." *Urban Studies* 51 (7): 1471–1486. https://doi.org/10.1177/0042098013500089.

Butt, Waqas H. 2019. "Beyond the Abject: Caste and the Organization of Work in Pakistan's Waste Economy." *International Labor and Working-Class History* 95:18–33. https://doi.org/10.1017/S0147547919000061.

Cairns, Kate, Kim de Laat, Josée Johnston, and Shyon Baumann. 2020. "The Caring, Committed Eco-Mom: Consumption Ideals and Lived Realities of Toronto Mothers." In *Green Consumption*, edited by Bart Barendregt and Rivke Jaffe, 100–114. Abingdon, UK: Routledge.

Calleja, Olivia. 2021. "Contingent Resistance: The Politics of Waste Commons in Neoliberal Delhi." *Contributions to Indian Sociology* 55 (3): 416–437. https://doi.org/10.1177/00699667211059715.

Cardullo, Paolo, and Rob Kitchin. 2019. "Smart Urbanism and Smart Citizenship: The Neoliberal Logic of 'Citizen-Focused' Smart Cities in Europe." *Environment and Planning C: Politics and Space* 37 (5): 813–830. https://doi.org/10.1177/0263774X18806508.

Carfagna, Lindsey B., Emilie A. Dubois, Connor Fitzmaurice, Monique Y. Ouimette, Juliet B. Schor, Margaret Willis, and Thomas Laidley. 2014. "An Emerging Eco-Habitus: The Reconfiguration of High Cultural Capital Practices among Ethical Consumers." *Journal of Consumer Culture* 14 (2): 158–178. https://doi.org/10.1177/1469540514526227.

Carré, Françoise, Pat Horn, and Chris Bonner. 2018. "Collective Bargaining by Informal Workers in the Global South: Where and How It Takes Place." WIEGO Working Paper 38. Women in Informal Employment: Globalizing and Organizing. https://www.wiego.org/publications/collective-bargaining-informal-workers-global-south-where-and-how-it-takes-place.

Castán Broto, Vanesa, and Susana Neves Alves. 2018. "Intersectionality Challenges for the Co-Production of Urban Services: Notes for a Theoretical and Methodological Agenda." *Environment and Urbanization* 30 (2): 367–386. https://doi.org/10.1177/0956247818790208.

Castán Broto, Vanesa, and Martín Sanzana Calvet. 2020. "Sacrifice Zones and the Construction of Urban Energy Landscapes in Concepción, Chile." *Journal of Political Ecology* 27 (1): 279–299. https://doi.org/10.2458/v27i1.23059.

Castán Broto, Vanesa, and Linda Westman. 2019. *Urban Sustainability and Justice: Just Sustainabilities and Environmental Planning.* London: Zed Books.

Catney, Philip, Sherilyn MacGregor, Andrew Dobson, Sarah Marie Hall, Sarah Royston, Zoe Robinson, Mark Ormerod, and Simon Ross. 2014. "Big Society, Little Justice? Community Renewable Energy and the Politics of Localism." *Local Environment* 19 (7): 715–730. https://doi.org/10.1080/13549839.2013.792044.

Chacko, Elizabeth. 2007. "From Brain Drain to Brain Gain: Reverse Migration to Bangalore and Hyderabad, India's Globalizing High Tech Cities." *GeoJournal* 68 (2): 131–140. https://doi.org/10.1007/s10708-007-9078-8.

Chacko, Priya. 2018. "The Right Turn in India: Authoritarianism, Populism and Neoliberalisation." *Journal of Contemporary Asia* 48 (4): 541–565. https://doi.org/10.1080/00472336.2018.1446546.

Chakrabarty, Dipesh. 1991. "Open Space/Public Place: Garbage, Modernity and India." *South Asia: Journal of South Asian Studies* 14 (1): 15–31. https://doi.org/10.1080/00856409108723146.

Chakravarty, Shoibal, and M. V. Ramana. 2012. "The Hiding behind the Poor Debate: A Synthetic Overview." In *Handbook of Climate Change and India: Development, Politics and Governance*, edited by Navroz K. Dubash, 218–227. New Delhi: Oxford University Press.

Chalfin, Brenda. 2014. "Public Things, Excremental Politics, and the Infrastructure of Bare Life in Ghana's City of Tema." *American Ethnologist* 41 (1): 92–109. https://doi.org/10.1111/amet.12062.

Chancel, Lucas, and Thomas Piketty. 2015. "Carbon and Inequality: From Kyoto to Paris: Trends in the Global Inequality of Carbon Emissions (1998–2013) and Prospects for an Equitable Adaptation Fund." Paris: Paris School of Economics. http://dx.doi.org/10.13140/RG.2.1.3536.0082.

Chandran, Pinky. 2021. "How the Pandemic Locked Waste Workers Out of Livelihoods." *Citizen Matters, Bengaluru* (blog). July 1, 2021. https://bengaluru.citizenmatters.in/pandemic-waste-workers-livelihoods-63472.

Chandran, Pinky, Kabir Arora, Marwan Abubaker, and Nalini Shekar. 2019. *Valuing Urban Waste: The Need for Comprehensive Material Recovery and Recycling Policy*. India: Unpublished report.

Chandran, Pinky, Sandya Narayanan, and Umashankar Subramanian. 2019. "A Mirage: Assessment of Swachh Bharat Abhiyan and SWM Rules 2016: Waste Pickers Perspective across India, September 2019." The Alliance of Indian Waste-Pickers. https://wasteframes.com/a-mirage-assessment-of-swachh-bharat-abhiyan-and-swm-rules-2016-wastepickers-perspective-across-india.

Chatterjee, Partha. 2006. *The Politics of the Governed: Reflections on Popular Politics in Most of the World*. Leonard Hastings Schoff Lectures edition. New York: Columbia University Press.

Chaturvedi, Bharati. 2009. "Cooling Agents: The Impact on the Informal Recycling Sector on Carbon Emissions." Delhi: Chintan-Environmental. www.chintan-india .org/documents/research_and_reports/chintan_ report_cooling_agents.pdf.

Chaturvedi, Bharati. 2014. "A Waste of Wealth: How Indian Cities Are Ignoring the Recyclers but Asking for Recycling." *Environmental Justice* 7 (5): 138–141. https://doi .org/10.1089/env.2014.0025.

Chaturvedi, Bharati, and Vinay Gidwani. 2011. "The Right to Waste: Informal Sector Recyclers and Struggles for Social Justice in Post-Reform Urban India." In *India's New Economic Policy: A Critical Analysis*, edited by Amitabh Kundu, Richard Peet, and Waquar Ahmed, 125–153. Abingdon, UK: Routledge.

Checker, Melissa. 2011. "Wiped Out by the 'Greenwave': Environmental Gentrification and the Paradoxical Politics of Urban Sustainability." *City and Society* 23 (2): 210–229. https://doi.org/10.1111/j.1548-744X.2011.01063.x.

Chen, Martha, Françoise Carré, and Françoise Carré. 2020. *The Informal Economy Revisited: Examining the Past, Envisioning the Future*. Abingdon, UK: Routledge.

CHF International and Mythri Sarva Seva Samithi. 2010. "Sample Study of Informal Waste-Pickers in Bangalore (September–October 2010)." https://pdfslide.net /documents/informal-waste-workers-contribution-bangalore-are-several-rules-and -national.html.

Chikarmane, Poornima, and Laxmi Narayan. 2005. "Organising the Unorganised: A Case Study of the Kagad Kach Patra Kashtakari Panchayat (Trade Union of Waste-Pickers)." Pune: KKPKP. https://swachcoop.com/pdf/casestudy-kagadkachpatrackashtakari.pdf.

Chini, Christopher M., James F. Canning, Kelsey L. Schreiber, Joshua M. Peschel, and Ashlynn S. Stillwell. 2017. "The Green Experiment: Cities, Green Stormwater Infrastructure, and Sustainability." *Sustainability* 9 (1): 105. https://doi.org/10.3390 /su9010105.

Chintan. 2012. "Give Back Our Waste: What the Okhla Waste-to-Energy Plant Has Done to Local Wastepickers." Chintan Environmental Research and Action Group. http://www.chintan-india.org/documents/research_and_reports/chintan-report-give -back-our-waste.pdf.

Chu, Eric, and Kavya Michael. 2019. "Recognition in Urban Climate Justice: Marginality and Exclusion of Migrants in Indian Cities." *Environment and Urbanization* 31 (1): 139–156. https://doi.org/10.1177/0956247818814449.

Coca-Cola Company. n.d. "Moving toward a Circular Economy—News & Articles." Coca-Cola Company. Accessed November 21, 2021. https://www.coca-colacompany .com/news/moving-toward-a-circular-economy.

Coelho, Karen. 2020. "Lines in the Mud: Tank Eco-Restoration and Boundary Contestations in Chennai." *Urbanisation* 5 (2): 121–139. https://doi.org/10.1177 /2455747120965508.

Coelho, Karen, Lalitha Kamath, and M. Vijayabaskar. 2020. "Opening Up or Ushering In? Citizen Participation as Mandate and Practice in Urban Governance." In *Participolis: Consent and Contention in Neoliberal Urban India*, edited by Karen Coelho, Lalitha Kamath, and M. Vijayabaskar, 3–33. New Delhi: Routledge India.

Coelho, Karen, and T. Venkat. 2009. "The Politics of Civil Society: Neighbourhood Associationism in Chennai." *Economic and Political Weekly* 44 (26/27): 358–367. https://www.jstor.org/stable/40279795.

Cohen, Daniel Aldana. 2017. "The Other Low-Carbon Protagonists: Poor People's Movements and Climate Politics in São Paulo." Chapter 7 in *The City Is the Factory*, edited by Miriam Greenberg and Penny Lewis, 140–157. Ithaca, NY: Cornell University Press. https://doi.org/10.7591/9781501708060-010.

Collins, Patricia Hill. 2002. *Black Feminist Thought: Knowledge, Consciousness, and the Politics of Empowerment*. Abingdon, UK: Routledge.

Collins, Randall. 2014. "Interaction Ritual Chains and Collective Effervescence." In *Collective Emotions*, edited by Christian Von Scheve, and Mikko Salmela, 299–311. Oxford: Oxford University Press.

Corvellec, Hervé, María José Zapata Campos, and Patrik Zapata. 2013. "Infrastructures, Lock-In, and Sustainable Urban Development: The Case of Waste Incineration in the Göteborg Metropolitan Area." *Journal of Cleaner Production* 50: 32–39. https://doi.org/10.1016/j.jclepro.2012.12.009.

Corvellec, Hervé, Johan Hultman, Anne Jerneck, Susanne Arvidsson, Johan Ekroos, Niklas Wahlberg, and Timothy W. Luke. 2021. "Resourcification: A Non-Essentialist Theory of Resources for Sustainable Development." *Sustainable Development* 29 (6): 1249–1256. https://doi.org/10.1002/sd.2222.

Corvellec, Hervé, Alison F. Stowell, and Nils Johansson. 2022. "Critiques of the Circular Economy." *Journal of Industrial Ecology* 26 (2): 421–432. https://doi.org/10.1111/jiec.13187.

Corwin, Julia. 2020. "Between Toxics and Gold: Devaluing Informal Labor in the Global Urban Mine." *Capitalism Nature Socialism* 31 (4): 106–123. https://doi.org/10.1080/10455752.2019.1690533.

Crenshaw, Kimberlé W. 2017. *On Intersectionality: Essential Writings*. New York: The New Press.

Creutzig, F., J. Roy, W. F. Lamb, I. M. L. Azevedo, W. Bruine de Bruin, H. Dalkmann, O. Y. Edelenbosch, et al. 2018. "Towards Demand-Side Solutions for Mitigating Climate Change." *Nature Climate Change* 8 (April): 268–271. https://doi.org/10.1038/s41558-018-0121-1.

Das, Diganta. 2020. "In Pursuit of Being Smart? A Critical Analysis of India's Smart Cities Endeavor." *Urban Geography* 41 (1): 55–78. https://doi.org/10.1080/02723638.2019.1646049.

Dauvergne, Peter. 2010. *The Shadows of Consumption: Consequences for the Global Environment*. Cambridge, MA: MIT Press.

Dauvergne, Peter. 2016. *Environmentalism of the Rich*. Cambridge, MA: MIT Press.

Deccan Herald. 2012. "Mafia Obstructing Scientific Disposal of Garbage in City." August 15, 2012. https://www.deccanherald.com/content/271701/mafia-obstructing -scientific-disposal-garbage.html.

Deccan Herald. 2013. "Mandur Villagers Set Garbage on Fire." February 1, 2013. http:// www.deccanherald.com/content/309314/mandur-villagers-set-garbage-fire.html.

DeFilippis, James, Robert Fisher, and Eric Shragge. 2006. "Neither Romance nor Regulation: Re-Evaluating Community." *International Journal of Urban and Regional Research* 30 (3): 673–689. https://doi.org/10.1111/j.1468-2427.2006.00680.x.

Demaria, Federico, and Seth Schindler. 2016. "Contesting Urban Metabolism: Struggles over Waste-to-Energy in Delhi, India." *Antipode* 48 (2): 293–313. https://doi .org/10.1111/anti.12191.

Demaria, Federico, and Marcos Todt. 2020. "How Waste Pickers in the Global South Are Being Sidelined by New Policies." *The Conversation*. http://theconversation.com /how-waste-pickers-in-the-global-south-are-being-sidelined-by-new-policies-132521.

De Souza, Rebecca T. 2019. *Feeding the Other: Whiteness, Privilege, and Neoliberal Stigma in Food Pantries*. Cambridge, MA: MIT Press.

Dhananka, Swetha Rao. 2010. "Claim-Making of Urban Poor and the Role of Inter-mediary Brokering Organisations." In *Boundaries: Dichotomies of Keeping In and Keeping Out*, edited by Julian Chapple, 43–57. Leiden: Brill.

Dias, Sonia Maria. 2016. "Waste Pickers and Cities." *Environment and Urbanization* 28 (2): 375–390. https://doi.org/10.1177/0956247816657302.

Dias, Sonia Maria, and Melanie Samson. 2016. *Informal Economy Monitoring Study Sector Report: Waste Pickers*. Cambridge, MA: WIEGO.

Di Chiro, Giovanna. 2008. "Living Environmentalisms: Coalition Politics, Social Reproduction, and Environmental Justice." *Environmental Politics* 17 (2): 276–298. https://doi.org/10.1080/09644010801936230.

Dinler, Demet Ş. 2016. "New Forms of Wage Labour and Struggle in the Informal Sector: The Case of Waste Pickers in Turkey." *Third World Quarterly* 37 (10): 1834–1854. https://doi.org/10.1080/01436597.2016.1175934.

Dobson, Andrew. 2006. "Ecological Citizenship: A Defence." *Environmental Politics* 15 (3): 447–451. https://doi.org/10.1080/09644010600627766.

Doherty, Jacob. 2021. *Waste Worlds: Inhabiting Kampala's Infrastructures of Disposability*. Berkeley: University of California Press.

Doherty, Jacob, and Kate Brown. 2019. "Labor Laid Waste: An Introduction to the Special Issue on Waste Work." *International Labor and Working-Class History* 95:1–17. https://doi.org/10.1017/S0147547919000048.

Doron, Assa. 2016. "Unclean, Unseen: Social Media, Civic Action and Urban Hygiene in India." *South Asia: Journal of South Asian Studies* 39 (4): 715–739. https://doi.org/10.1080/00856401.2016.1218096.

Doron, Assa, and Robin Jeffrey. 2018. *Waste of a Nation: Garbage and Growth in India.* Cambridge, MA: Harvard University Press.

Doshi, Sapana. 2013. "The Politics of the Evicted: Redevelopment, Subjectivity, and Difference in Mumbai's Slum Frontier." *Antipode* 45 (4): 844–865. https://doi.org/10.1111/j.1467-8330.2012.01023.x.

Doshi, Sapana. 2019. "Greening Displacements, Displacing Green: Environmental Subjectivity, Slum Clearance, and the Embodied Political Ecologies of Dispossession in Mumbai." *International Journal of Urban and Regional Research* 43 (1): 112–132. https://doi.org/10.1111/1468-2427.12699.

Doshi, Sapana, and Malini Ranganathan. 2017. "Contesting the Unethical City: Land Dispossession and Corruption Narratives in Urban India." *Annals of the American Association of Geographers* 107 (1): 183–199. https://doi.org/10.1080/24694452.2016.1226124.

Doshi, Sapana, and Malini Ranganathan. 2019. "Towards a Critical Geography of Corruption and Power in Late Capitalism." *Progress in Human Geography* 43 (3): 436–457. https://doi.org/10.1177/0309132517753070.

Douglas, Mary. 2003. *Purity and Danger: An Analysis of Concepts of Pollution and Taboo.* Abingdon, UK: Routledge.

Dubash, Navroz K. 2013. "The Politics of Climate Change in India: Narratives of Equity and Cobenefits." *WIREs Climate Change* 4 (3): 191–201. https://doi.org/10.1002/wcc.210.

Dubois, Ghislain, Benjamin Sovacool, Carlo Aall, Maria Nilsson, Carine Barbier, Alina Herrmann, Sébastien Bruyère, et al. 2019. "It Starts at Home? Climate Policies Targeting Household Consumption and Behavioral Decisions Are Key to Low-Carbon Futures." *Energy Research and Social Science* 52 (June): 144–158. https://doi.org/10.1016/j.erss.2019.02.001.

Du Bois, William Edward Burghardt. 1935. *Black Reconstruction in America: An Essay toward a History of the Part which Black Folk Played in the Attempt to Reconstruct Democracy in America, 1860–1880.* Oxford: Oxford University Press.

Durose, Catherine, and Liz Richardson. 2015. *Designing Public Policy for Co-production: Theory, Practice and Change.* Bristol, UK: Policy Press.

Dürr, Eveline, and Rivke Jaffe. 2010. *Urban Pollution: Cultural Meanings, Social Practices*. New York: Berghahn Books.

EJOLT. n.d. "Mavallipura Landfill Site at Bangalore, Karnataka, India | EJAtlas." Environmental Justice Atlas. Accessed May 17, 2021. https://ejatlas.org/conflict/mavallipura -landfill-site-bangalore-karnataka-india.

Ellen MacArthur Foundation. 2016. "Circular Economy in India: Rethinking Growth for Long-Term Prosperity." https://ellenmacarthurfoundation.org/circular-economy-in -india.

Ellis, Rowan. 2011. "Who's Participation? Who's Sustainability? A Critical Analysis of Initiatives for Urban Sustainability in India." *Scottish Geographical Journal* 127 (3): 193–208. https://doi.org/10.1080/14702541.2011.616863.

Ellis, Rowan. 2012. "'A World Class City of Your Own!': Civic Governmentality in Chennai, India." *Antipode* 44 (4): 1143–1160. https://doi.org/10.1111/j.1467-8330 .2011.00958.x.

Elwood, Sarah, Victoria Lawson, and Eric Sheppard. 2017. "Geographical Relational Poverty Studies." *Progress in Human Geography* 41 (6): 745–765. https://doi.org/10.1177 /0309132516659706.

Environment Support Group. 2010. "Bangalore's Toxic Legacy: Investigating Mavallipura's Illegal Landfills." http://www.esgindia.org/campaigns/mavallipura/press /bangalores-toxic-legacy-investigating-ma.html-0.

Erler, Mirka, and Christoph Dittrich. 2020. "Middle Class, Tradition and the Desi-Realm—Discourses of Alternative Food Networks in Bengaluru, India." *Sustainability* 12 (7): 2741. https://doi.org/10.3390/su12072741.

Erler, Mirka, Markus Keck, and Christoph Dittrich. 2020. "The Changing Meaning of Millets: Organic Shops and Distinctive Consumption Practices in Bengaluru, India." *Journal of Consumer Culture*, 22 (1): 124–142. https://doi.org/10.1177/14695405209 02508.

European Commission. 2017. "Study on the Review of the List of Critical Raw Materials: Critical Raw Materials Factsheets." ET-04-15-307-EN-N. Brussels: European Commission. http://publications.europa.eu/resource/cellar/7345e3e8-98fc-11e7-b92d -01aa75ed71a1.0001.01/DOC_1.

European Commission. 2018. "Circular Economy and Resource Efficiency: Potential and Opportunities for Transformative Global Climate Action." Luxembourg: Publications Office of the European Union. https://ec.europa.eu/environment/international _issues/pdf/COP24%20EU%20Side%20Event%20(4%20December%202018,%20 Katowice,%20Poland)%20-%20report.pdf.

Express News Service. 2022. "Regularise All 33k Sanitation Workers: Pourakarmika Forums." September 21, 2022. https://www.newindianexpress.com/states/karnataka

/2022/sep/21/regularise-all-33k-sanitation-workers-pourakarmika-forums-2500342.html.

Fahmi, Wael Salah, and Keith Sutton. 2006. "Cairo's Zabaleen Garbage Recyclers: Multi-Nationals' Takeover and State Relocation Plans." *Habitat International* 30 (4): 809–837. https://doi.org/10.1016/j.habitatint.2005.09.006.

Farbotko, Carol. 2018. *Domestic Environmental Labour: An Ecofeminist Perspective on Making Homes Greener.* Abingdon, UK: Routledge.

Federici, Silvia. 2008. "Precarious Labor: A Feminist Viewpoint." *In the Middle of a Whirlwind* (blog). June 6, 2008. https://inthemiddleofthewhirlwind.wordpress.com/precarious-labor-a-feminist-viewpoint.

Ferguson, James. 1994. *The Anti-Politics Machine: Development, Depoliticization, and Bureaucratic Power in Lesotho.* Minneapolis: University of Minnesota Press.

Fergutz, Oscar, Sonia Dias, and Diana Mitlin. 2011. "Developing Urban Waste Management in Brazil with Waste Picker Organizations." *Environment and Urbanization* 23 (2): 597–608. https://doi.org/10.1177/0956247811418742.

Fernandes, Leela. 2009. "The Political Economy of Lifestyle: Consumption, India's New Middle Class and State-Led Development." In *The New Middle Classes*, edited by Lars Meier and Hellmuth Lange, 219–236. Dordrecht: Springer Netherlands.

Fernandes, Leela, and Patrick Heller. 2006. "Hegemonic Aspirations." *Critical Asian Studies* 38 (4): 495–522. https://doi.org/10.1080/14672710601073028.

Forno, Francesca, and Paolo R. Graziano. 2014. "Sustainable Community Movement Organisations." *Journal of Consumer Culture* 14 (2): 139–157. http://dx.doi.org/10.1177/1469540514526225.

Fraser, Nancy. 2021. "Climates of Capital for a Trans-Environmental Eco-Socialism." *New Left Review* 127: 94–127. https://newleftreview.org/issues/ii127/articles/nancy-fraser-climates-of-capital.pdf.

Frazier, Camille. 2018. "'Grow What You Eat, Eat What You Grow': Urban Agriculture as Middle Class Intervention in India." *Journal of Political Ecology* 25 (1): 221–238. https://doi.org/10.2458/v25i1.22970.

Fredericks, Rosalind. 2018. *Garbage Citizenship: Vital Infrastructures of Labor in Dakar, Senegal.* Durham, NC: Duke University Press.

Fredericks, Rosalind. 2022. "Anthropocenic Discards: Embodied Infrastructures and Uncanny Exposures at Dakar's Dump." *Antipode*, December 13. https://doi.org/10.1111/anti.12796.

Furlong, Kathryn. 2011. "Small Technologies, Big Change: Rethinking Infrastructure through STS and Geography." *Progress in Human Geography* 35 (4): 460–482. https://doi.org/10.1177/0309132510380488.

Ganguly, Sunayana. 2017. "Making Sustainability Palatable? Changing Practices of Middle-Class Food Consumption in Bangalore." *International Development Policy | Revue Internationale de Politique de Développement* 8, no. 2 (September). https://doi.org /10.4000/poldev.2478.

Ghertner, Asher. 2011a. "Rule by Aesthetics: World-Class City Making in Delhi." In *Worlding Cities: Asian Experiments and the Art of Being Global*, edited by Ananya Roy and Aihwa Ong, 279–306. Chichester, UK: Wiley-Blackwell. http://onlinelibrary .wiley.com/doi/10.1002/9781444346800.ch11/summary.

Ghertner, Asher. 2011b. "Gentrifying the State, Gentrifying Participation: Elite Governance Programs in Delhi." *International Journal of Urban and Regional Research* 35 (3): 504–532. https://doi.org/10.1111/j.1468-2427.2011.01043.x.

Ghertner, Asher. 2012. "Nuisance Talk and the Propriety of Property: Middle Class Discourses of a Slum-Free Delhi." *Antipode* 44 (4): 1161–1187. https://doi.org/10.1111 /j.1467-8330.2011.00956.x.

Ghertner, Asher. 2015. *Rule by Aesthetics: World-Class City Making in Delhi*. New York: Oxford University Press.

Ghosh, Asha. 2005. "Public-Private or a Private Public? Promised Partnership of the Bangalore Agenda Task Force." *Economic and Political Weekly* 40 (47): 4914–4922. https://www.jstor.org/stable/4417426.

Ghosh, Bipashyee, and Saurabh Arora. 2022. "Smart as (Un)Democratic? The Making of a Smart City Imaginary in Kolkata, India." *Environment and Planning C: Politics and Space* 40 (1): 318–339. https://doi.org/10.1177/23996544211027583.

Gibson-Graham, J. K. 2006. *The End of Capitalism (as We Knew It): A Feminist Critique of Political Economy*. Minneapolis: University of Minnesota Press.

Gibson-Graham, J. K. 2008. "Diverse Economies: Performative Practices for 'Other Worlds.'" *Progress in Human Geography* 32 (5): 613–632. https://doi.org/10.1177 /0309132508090821.

Gidwani, Vinay. 2013a. "Value Struggles: Waste Work and Urban Ecology." In *Ecologies of Urbanism in India: Metropolitan Civility and Sustainability*, edited by K. Sivaramakrishnan and Anne Rademacher, 169–200. Hong Kong: Hong Kong University Press.

Gidwani, Vinay. 2013b. "Six Theses on Waste, Value, and Commons." *Social and Cultural Geography* 14 (7): 773–783. https://doi.org/10.1080/14649365.2013.800222.

Gidwani, Vinay. 2015. "The Work of Waste: Inside India's Infra-Economy." *Transactions of the Institute of British Geographers* 40 (4): 575. https://doi.org/10.1111/tran.12094.

Gidwani, Vinay, and Bharati Chaturvedi. 2013. "Poverty as Geography: Motility, Stoppage and Circuits of Waste in Delhi." In *Urban Navigations: Politics, Space and the*

City in South Asia, edited by Jonathan Shapiro Anjaria and Colin McFarlane, 50–78. New Delhi: Routledge India.

Gidwani, Vinay, and Anant Maringanti. 2016. "The Waste-Value Dialectic: Lumpen Urbanization in Contemporary India." *Comparative Studies of South Asia, Africa and the Middle East* 36 (1): 112–133. https://doi.org/10.1215/1089201x -3482159.

Gidwani, Vinay, and Rajyashree N. Reddy. 2011. "The Afterlives of 'Waste': Notes from India for a Minor History of Capitalist Surplus." *Antipode* 43 (5): 1625–1658. https://doi.org/10.1111/j.1467-8330.2011.00902.x.

Gill, Kaveri. 2007. "Interlinked Contracts and Social Power: Patronage and Exploitation in India's Waste Recovery Market." *Journal of Development Studies* 43 (8): 1448–1474. https://doi.org/10.1080/00220380701611519.

Gill, Kaveri. 2009. *Of Poverty and Plastic: Scavenging and Scrap Trading Entrepreneurs in India's Urban Informal Economy.* Oxford: Oxford University Press.

Gill, Kaveri. 2021. "The Environment as Disingenuous Trope: Tracing Waste Policy and Practice in a Medium Hill Town of the Himalayas, India." *Journal of Developing Societies* 37 (2): 184–200. https://doi.org/10.1177/0169796X211001246.

Glassman, Jim. 2016. "Primitive Accumulation, Accumulation by Dispossession, Accumulation by 'Extra-Economic' Means." *Progress in Human Geography* 30(5): 608–625. https://doi.org/10.1177/0309132506070172.

Glenn, Evelyn Nakano. 2015. "Settler Colonialism as Structure: A Framework for Comparative Studies of US Race and Gender Formation." *Sociology of Race and Ethnicity* 1 (1): 52–72. https://doi.org/10.1177/2332649214560440.

global_rec. 2018. "Indore Waste Management Model Is Incomplete without Participation of Waste-Pickers." *Global Alliance of Waste Pickers* (blog). August 30, 2018. https://globalrec.org/2018/08/30/indore-waste-management-model-is-incomplete -without-participation-of-waste-pickers.

global_rec. 2022. "Call for Recognition of Waste-Pickers in International Legally Binding Instrument on Plastics Pollution." *Global Alliance of Waste Pickers* (blog). February 22, 2022. https://globalrec.org/2022/02/22/declaration-wastepickers-inclusion-unea.

Goldman, Michael. 2011. "Speculative Urbanism and the Making of the Next World City." *International Journal of Urban and Regional Research* 35 (3): 555–581. https://doi .org/10.1111/j.1468-2427.2010.01001.x.

Gopakumar, Govind. 2020. *Installing Automobility: Emerging Politics of Mobility and Streets in Indian Cities.* Cambridge, MA: MIT Press.

Gould, Kenneth, and Tammy Lewis. 2016. *Green Gentrification: Urban Sustainability and the Struggle for Environmental Justice.* London: Routledge.

Gower, Richard, and Patrick Schröder. 2016. "Virtuous Circle: How the Circular Economy Can Create Jobs and Save Lives in Low and Middle-Income Countries." IDS/Tearfund. https://www.researchgate.net/publication/306562812_Virtuous_Circle_how_the_circular_economy_can_create_jobs_and_save_lives_in_low_and_middle-income_countries.

Graham, Stephen, and Colin McFarlane. 2014. *Infrastructural Lives: Urban Infrastructure in Context*. Abingdon, UK: Routledge.

Green, Gary Paul, and Anna Haines. 2015. *Asset Building and Community Development*. Newbury Park, CA: Sage.

Greenberg, Miriam. 2013. "What on Earth Is Sustainable?: Toward Critical Sustainability Studies." *Boom: A Journal of California* 3 (4): 54–66. https://doi.org/10.1525/boom.2013.3.4.54.

Gregson, Nicky, Mike Crang, Julie Botticello, Melania Calestani, and Anna Krzywoszynska. 2016. "Doing the 'Dirty Work' of the Green Economy: Resource Recovery and Migrant Labour in the EU." *European Urban and Regional Studies* 23 (4): 541–555. https://doi.org/10.1177/0969776414554489.

Gregson, Nicky, Mike Crang, Sara Fuller, and Helen Holmes. 2015. "Interrogating the Circular Economy: The Moral Economy of Resource Recovery in the EU." *Economy and Society* 44 (2): 218–243. https://doi.org/10.1080/03085147.2015.1013353.

GRID_Arendal. 2022. "A Seat at the Table—The Role of the Informal Recycling Sector in Plastic Pollution Reduction, and Recommended Policy Changes | GRID-Arendal." https://www.grida.no/publications/863.

Grossmann, Mena, and Emily Creamer. 2017. "Assessing Diversity and Inclusivity within the Transition Movement: An Urban Case Study." *Environmental Politics* 26 (1): 161–182. https://doi.org/10.1080/09644016.2016.1232522.

Guha, Ramachandra. 1989. "Radical American Environmentalism and Wilderness Perservation: A Third World Critique." *Environmental Ethics* 11 (1): 71–83. https://doi.org/10.5840/enviroethics198911123.

Guha-Khasnobis, Basudeb, and Ravi Kanbur. 2006. *Linking the Formal and Informal Economy: Concepts and Policies*. Oxford: Oxford University Press.

Guibrunet, Louise. 2019. "What Is 'Informal' in Informal Waste Management? Insights from the Case of Waste Collection in the Tepito Neighbourhood, Mexico City." *Waste Management* 86 (March): 13–22. https://doi.org/10.1016/j.wasman.2019.01.021.

Guibrunet, Louise. 2021. "The Reformist Sustainability Discourse and the Exclusion of the Informal Economy from Mexico City's Environmental Policies." *Local Environment* 26 (1): 1–16. https://doi.org/10.1080/13549839.2020.1861588.

Gullion, Jessica Smartt. 2015. *Fracking the Neighborhood: Reluctant Activists and Natural Gas Drilling*. Cambridge, MA: MIT Press.

Gupta, Hemangini. 2018. "In Bengaluru's Gated Communities, New Forms of Civil Engagement Are Emerging." *Economic and Political Weekly* 53 (39). https://www.epw .in/sites/default/files/engage_pdf/2018/10/01/152722.pdf.

Gupta, Hemangini. 2019. "Testing the Future: Gender and Technocapitalism in Start-Up India." *Feminist Review* 123 (1): 74–88. https://doi.org/10.1177/0141778919879740.

Gupta, Pallavi. 2022. "Broomscapes: Racial Capitalism, Waste, and Caste in Indian Railway Stations." *Ethnic and Racial Studies* 45 (2): 235–256. https://doi.org/10.1080 /01419870.2021.1964557.

Guru, Gopal, ed. 2011. *Humiliation: Claims and Context*. New Delhi: Oxford University Press.

Gutberlet, Jutta. 2012. *Recovering Resources—Recycling Citizenship: Urban Poverty Reduction in Latin America*. Farnham, UK: Ashgate.

Gutberlet, Jutta. 2021. "Grassroots Waste Picker Organizations Addressing the UN Sustainable Development Goals." *World Development* 138 (February): 105195. https://doi.org/10.1016/j.worlddev.2020.105195.

Gutberlet, Jutta, and Sebastián Carenzo. 2020. "Waste Pickers at the Heart of the Circular Economy: A Perspective of Inclusive Recycling from the Global South." *Worldwide Waste: Journal of Interdisciplinary Studies* 3 (1): 6. https://doi.org/10.5334/wwwj.50.

Gutberlet, Jutta, Sebastián Carenzo, Jaan-Henrik Kain, and Adalberto Mantovani Martiniano de Azevedo. 2017. "Waste Picker Organizations and Their Contribution to the Circular Economy: Two Case Studies from a Global South Perspective." *Resources* 6 (4): 52. https://doi.org/10.3390/resources6040052.

Gutberlet, Jutta, Jaan-Henrik Kain, Belinda Nyakinya, Dickens H. Ochieng, Nicholas Odhiambo, Michael Oloko, John Omolo, et al. 2016. "Socio-Environmental Entrepreneurship and the Provision of Critical Services in Informal Settlements." *Environment and Urbanization* 28 (1): 205–222. https://doi.org/10.1177/0956247815623772.

Gutberlet, Jutta, Santiago Sorroche, Angela Martins Baeder, Patrik Zapata, and María José Zapata Campos. 2021. "Waste Pickers and Their Practices of Insurgency and Environmental Stewardship." *Journal of Environment and Development* 30 (4): 369–394. https://doi.org/10.1177/10704965211055328.

Guthman, Julie. 2003. "Fast Food/Organic Food: Reflexive Tastes and the Making of 'Yuppie Chow.'" *Social and Cultural Geography* 4 (1): 45–58. https://doi.org/10.1080 /1464936032000049306.

Guthman, Julie. 2008. "'If They Only Knew': Color Blindness and Universalism in California Alternative Food Institutions." *Professional Geographer* 60 (3): 387–397. https://doi.org/10.1080/00330120802013679.

Hall, Sarah Marie. 2020. "Social Reproduction as Social Infrastructure." *Soundings: A Journal of Politics and Culture* 76 (1): 82–94. https://doi.org/10.3898/SOUN.76.06.2020.

Harding, Sandra G. 2004. *The Feminist Standpoint Theory Reader: Intellectual and Political Controversies*. London: Psychology Press.

Hardt, Michael, and Antonio Negri. 2000. *Empire*. Cambridge, MA: Harvard University Press.

Harris, Gardiner. 2012. "India's Plague, Trash, Drowns Bangalore, Its Garden City." *New York Times*, October 26, 2012. http://www.nytimes.com/2012/10/27/world/asia /indias-plague-trash-drowns-bangalore-its-garden-city.html.

Harriss, John. 2006. "Middle-Class Activism and the Politics of the Informal Working Class." *Critical Asian Studies* 38 (4): 445–465. https://doi.org/10.1080/14672710 601073002.

Harriss-White, Barbara. 2017. "Matter in Motion: Work and Livelihoods in India's Economy of Waste." In *Critical Perspectives on Work and Employment in Globalizing India*, edited by Ernesto Noronha and Premilla D'Cruz, 95–111. Singapore: Springer Singapore. https://doi.org/10.1007/978-981-10-3491-6_6.

Hartmann, Chris, Christine Hegel, and Owusu Boampong. 2022. "The Forgotten Essential Workers in the Circular Economy? Waste Picker Precarity and Resilience amidst the COVID-19 Pandemic." *Local Environment* 27 (10–11): 1272–1286. https:// doi.org/10.1080/13549839.2022.2040464.

Harvey, David. 2010. "The Right to the City: From Capital Surplus to Accumulation by Dispossession." In *Accumulation by Dispossession*, edited by Swapna Banerjee-Guja, 17–32. New Delhi: Sage.

Hawkins, Gay. 2001. "Plastic Bags: Living with Rubbish." *International Journal of Cultural Studies* 4 (1): 5–23. https://doi.org/10.1177/136787790100400101.

Hébert, Karen, and Diana Mincyte. 2014. "Self-Reliance beyond Neoliberalism: Rethinking Autonomy at the Edges of Empire." *Environment and Planning D: Society and Space* 32 (2): 206–222. https://doi.org/10.1068/d6312.

Heiges, Jessica, and Kate O'Neill. 2022. "A Recycling Reckoning: How Operation National Sword Catalyzed a Transition in the U.S. Plastics Recycling System." *Journal of Cleaner Production*, October, 134367. https://doi.org/10.1016/j.jclepro.2022 .134367.

Hesketh, Chris. 2022. "Clean Development or the Development of Dispossession? The Political Economy of Wind Parks in Southern Mexico." *Environment and Planning E: Nature and Space* 5 (2): 543–565. https://doi.org/10.1177/2514848621991764.

Hickel, Jason. 2017. *The Divide: A Brief Guide to Global Inequality and Its Solutions*. New York: Random House.

Hickel, Jason. 2020. *Less Is More: How Degrowth Will Save the World*. New York: Random House.

Hickel, Jason, Daniel W. O'Neill, Andrew L. Fanning, and Huzaifa Zoomkawala. 2022. "National Responsibility for Ecological Breakdown: A Fair-Shares Assessment of Resource Use, 1970–2017." *Lancet Planetary Health* 6 (4): e342–e349. https://doi .org/10.1016/S2542-5196(22)00044-4.

Hindu, The. 2016. "No One Wants to Live near a Dry Waste Centre." October 3, 2016, sec. Bengaluru. https://www.thehindu.com/news/cities/bangalore/No-one-wants -to-live-near-a-dry-waste-centre/article15424314.ece.

Hindu, The. 2021a. "Five DWCCs Gutted in Fires in 15 Days, Foul Play Suspected." February 12, 2021, sec. Bengaluru. https://www.thehindu.com/news/cities/bangalore /five-dwccs-gutted-in-fires-in-15-days-foul-play-suspected/article33813277.ece.

Hindu, The. 2021b. "Pourkarmikas Begin Jatha, Press for Long-Pending Demands." October 10, 2021, sec. Bengaluru. https://www.thehindu.com/news/cities/bangalore /pourkarmikas-begin-jatha-press-for-long-pending-demands/article36923112.ece.

Ho, Vivian. 2019. "Wealthy Opponents of New Shelter Claim Homeless Are Bad for Environment." *The Guardian*, July 11, 2019, sec. US news. https://www.theguardian .com/us-news/2019/jul/11/san-francisco-navigation-center-shelter-lawsuit.

Hodges, Sarah. 2013. "Medical Garbage and the Making of Neo-Liberalism in India." *Economic and Political Weekly* 48, no. 48 (November 2013): 112–119. https://www .jstor.org/stable/23528937.

Hodges, Sarah. 2018. "Plastic History, Caste and the Government of Things in Modern India." In *South Asian Governmentalities: Michel Foucault and the Question of Postcolonial Orderings*, edited by Deana Heath and Stephen Legg, 178–199. Cambridge, UK: Cambridge University Press. https://doi.org/10.1017/9781108571982.008.

Hodgson, Dorothy Louise. 2011. *Being Maasai, Becoming Indigenous: Postcolonial Politics in a Neoliberal World*. Bloomington: Indiana University Press.

Hodson, Mike, and Simon Marvin. 2014. *After Sustainable Cities?* London: Routledge.

Hoffmann, Melody L., and Adonia Lugo. 2014. "Who Is 'World Class'? Transportation Justice and Bicycle Policy." *Urbanities* 4 (1): 45–61. https://www.anthrojournal -urbanities.com/docs/tableofcontents_6/5-%20Hoffmann-Lugo.pdf.

hooks, bell. 1996. *Teaching to Transgress: Education as the Practice of Freedom*. London: Routledge.

hooks, bell. 2014. *Yearning: Race, Gender, and Cultural Politics*. Abingdon, UK: Routledge.

Hyams, Keith, and Tina Fawcett. 2013. "The Ethics of Carbon Offsetting." *WIREs Climate Change* 4 (2): 91–98. https://doi.org/10.1002/wcc.207.

ILO. 2018. "Women and Men in the Informal Economy: A Statistical Picture. Third Edition." Report. International Labour Organization. http://www.ilo.org/global/publi cations/books/WCMS_626831/lang--en/index.htm.

Isenhour, Cindy, Gary McDonogh, and Melissa Checker. 2014. *Sustainability in the Global City: Myth and Practice*. Cambridge, UK: Cambridge University Press.

Jack, Tullia, Manisha Anantharaman, and Alison L. Browne. 2020. "'Without Cleanliness We Can't Lead the Life, No?' Cleanliness Practices, (In)Accessible Infrastructures, Social (Im)Mobility and (Un)Sustainable Consumption in Mysore, India." *Social and Cultural Geography* 23 (6): 814–835. https://doi.org/10.1080/14649365.2020.1820561.

Jambeck, Jenna R., Roland Geyer, Chris Wilcox, Theodore R. Siegler, Miriam Perryman, Anthony Andrady, Ramani Narayan, and Kara Lavender Law. 2015. "Plastic Waste Inputs from Land into the Ocean." *Science* 347 (6223): 768–771. https://doi .org/10.1126/science.1260352.

Johnston, Josée. 2008. "The Citizen-Consumer Hybrid: Ideological Tensions and the Case of Whole Foods Market." *Theory and Society* 37 (3): 229–270. https://doi.org/10 .1007/s11186-007-9058-5.

Johnston, Josee, Michelle Szabo, and Alexandra Rodney. 2011. "Good Food, Good People: Understanding the Cultural Repertoire of Ethical Eating." *Journal of Consumer Culture* 11 (3): 293–318. https://doi.org/10.1177/1469540511417996.

Joseph, Miranda. 2002. *Against the Romance of Community*. Minneapolis: University of Minnesota Press.

Kaijser, Anna, and Annica Kronsell. 2014. "Climate Change through the Lens of Intersectionality." *Environmental Politics* 23 (3): 417–433. https://doi.org/10.1080/09644016 .2013.835203.

Kain, Jaan-Henrik, Patrik Zapata, Adalberto Azevedo, Sebastián Carenzo, Goodluck Charles, Jutta Gutberlet, Michael Oloko, Jessica Perez-Reynosa, and María José Zapata Campos. 2022. "Characteristics, Challenges and Innovations of Waste Picker Organizations: A Comparative Perspective between Latin American and East African Countries." *PLOS One* 17 (July): e0265889. https://doi.org/10.1371/journal.pone.0265889.

Kalappa, Bansy. 2022. "Karnataka: 13,133 Pourakarmikas Made Govt Staffers." *New Indian Express*, September 20, 2022. https://www.newindianexpress.com/states/karnataka /2022/sep/20/karnataka-13133-pourakarmikas-made-govt-staffers-2499985.html.

Kama, Kärg. 2015. "Circling the Economy: Resource-Making and Marketization in EU Electronic Waste Policy." *Area* 47 (1): 16–23. http://dx.doi.org/10.1111/area.12143.

Kamath, Lalitha, and M. Vijayabaskar. 2014. "Middle-Class and Slum-Based Collective Action in Bangalore Contestations and Convergences in a Time of Market Reforms." *Journal of South Asian Development* 9 (2): 147–171. https://doi.org/10.1177 /0973174114536098.

Kappan, Rasheed. 2021. "No Empathy for Informal Waste Pickers, Seen as Dirty: Study by BBC Media Action." *Deccan Herald*, April 15, 2021. https://www.deccanherald.com

/city/top-bengaluru-stories/no-empathy-for-informal-waste-pickers-seen-as-dirty-study-by-bbc-media-action-974707.html.

Kashwan, Prakash. 2020. "American Environmentalism's Racist Roots Have Shaped Global Thinking about Conservation." *The Conversation.* https://theconversation.com/american-environmentalisms-racist-roots-have-shaped-global-thinking-about-conservation-143783.

Kashwan, Prakash. 2022. "Globalization of Environmental Justice: A Framework for Comparative Research." In *The Oxford Handbook of Comparative Environmental Politics*, edited by Jeannie Sowers, Stacy D. VanDeveer, and Erika Weinthal. Oxford: Oxford University Press. https://doi.org/10.1093/oxfordhb/9780197515037.013.4.

Kashwan, P. 2023. "Social Mobilizations for Climate Action and Climate Justice in India." In *Climate Justice in India*, edited by Prakash Kashwan, 140–161.

Kashwan, Prakash, Praneeta Mudaliar, Sheila R. Foster, and Floriane Clement. 2021. "Reimagining and Governing the Commons in an Unequal World: A Critical Engagement." *Current Research in Environmental Sustainability* 3 (January): 100102. https://doi.org/10.1016/j.crsust.2021.100102.

Katz, Cindi. 2001. "Vagabond Capitalism and the Necessity of Social Reproduction." *Antipode* 33 (4): 709–728. https://doi.org/10.1111/1467-8330.00207.

Kaviraj, Sudipta. 1998. "Filth and the Public Sphere: Concepts and Practices about Space in Calcutta." *Public Culture* 10:83–113. https://doi.org/doi:10.1215/08992363-10-1-83.

Kaza, Silpa, Lisa C. Yao, Perinaz Bhada-Tata, and Frank Van Woerden. 2018. *What a Waste 2.0: A Global Snapshot of Solid Waste Management to 2050.* Washington, DC: World Bank. https://doi.org/10.1596/978-1-4648-1329-0.

Kelley, Robin D. G. 2017. "What Did Cedric Robinson Mean by Racial Capitalism." *Boston Review* 12 (January 12, 2017). https://www.bostonreview.net/articles/robin-d-g-kelley-introduction-race-capitalism-justice.

Kenis, Anneleen. 2016. "Ecological Citizenship and Democracy: Communitarian versus Agonistic Perspectives." *Environmental Politics* 25 (6): 949–970. https://doi.org/10.1080/09644016.2016.1203524.

Kennedy, Emily Huddart, and Christine Horne. 2020. "Accidental Environmentalist or Ethical Elite? The Moral Dimensions of Environmental Impact." *Poetics* 82:101448. https://doi.org/10.1016/j.poetic.2020.101448.

Kennedy, Emily Huddart, Josée Johnston, and John R. Parkins. 2018. "Small-p Politics: How Pleasurable, Convivial and Pragmatic Political Ideals Influence Engagement in Eat-Local Initiatives." *British Journal of Sociology* 69 (3): 670–690. https://doi.org/10.1111/1468-4446.12298.

Kenner, Dario. 2019. *Carbon Inequality: The Role of the Richest in Climate Change.* London: Routledge.

Knapp, Freyja L. 2016. "The Birth of the Flexible Mine: Changing Geographies of Mining and the e-Waste Commodity Frontier." *Environment and Planning A: Economy and Space* 48 (10): 1889–1909. https://doi.org/10.1177/0308518X16652398.

Kornberg, Dana. 2019a. "From Balmikis to Bengalis: The 'Casteification' of Muslims in Delhi's Informal Garbage Economy." *Economic and Political Weekly* 54 (47): 48–54. https://www.epw.in/journal/2019/47/review-urban-affairs/balmikis-bengalis.html.

Kornberg, Dana. 2019b. "Garbage as Fuel: Pursuing Incineration to Counter Stigma in Postcolonial Urban India." *Local Environment* 24 (1): 1–17. https://doi.org/10.1080/13549839.2018.1545752.

Kornberg, Dana. 2020. "Competing for Jurisdiction: Practical Legitimation and the Persistence of Informal Recycling in Urban India." *Social Forces* 99 (2): 797–819. https://doi.org/10.1093/sf/soz169.

Larner, Wendy. 2003. "Neoliberalism?" *Environment and Planning D: Society and Space* 21 (5): 509–512. https://doi.org/10.1068/d2105ed.

Lau, Winnie W. Y., Yonathan Shiran, Richard M. Bailey, Ed Cook, Martin R. Stuchtey, Julia Koskella, Costas A. Velis, et al. 2020. "Evaluating Scenarios toward Zero Plastic Pollution." *Science* 369 (6510): 1455–1461. https://doi.org/10.1126/science.aba9475.

Lawhon, Mary, Henrik Ernstson, and Jonathan Silver. 2014. "Provincializing Urban Political Ecology: Towards a Situated UPE through African Urbanism." *Antipode* 46 (2): 497–516. https://doi.org/10.1111/anti.12051.

Lawhon, Mary, David Nilsson, Jonathan Silver, Henrik Ernstson, and Shuaib Lwasa. 2018. "Thinking through Heterogeneous Infrastructure Configurations." *Urban Studies* 55 (4): 720–732. https://doi.org/10.1177/0042098017720149.

Lawson, Victoria, and Sarah Elwood. 2014. "Encountering Poverty: Space, Class, and Poverty Politics." *Antipode* 46 (1): 209–228. https://doi.org/10.1111/anti.12030.

Le Billon, Philippe, and Rosaleen V. Duffy. 2018. "Conflict Ecologies: Connecting Political Ecology and Peace and Conflict Studies." *Journal of Political Ecology* 25 (1): 239–260. https://doi.org/10.2458/v25i1.22704.

Lee, Jemyung, Oliver Taherzadeh, and Keiichiro Kanemoto. 2021. "The Scale and Drivers of Carbon Footprints in Households, Cities and Regions across India." *Global Environmental Change* 66 (January): 102205. https://doi.org/10.1016/j.gloenvcha.2020.102205.

Lee, Joel. 2017. "Odor and Order: How Caste Is Inscribed in Space and Sensoria." *Comparative Studies of South Asia, Africa and the Middle East* 37 (3): 470–490. https://muse.jhu.edu/article/683279.

Lele, Sharachchandra, Eduardo S. Brondizio, John Byrne, Georgina M. Mace, and Joan Martinez-Alier. 2019. *Rethinking Environmentalism: Linking Justice, Sustainability, and Diversity*. Cambridge, MA: MIT Press.

Lepawsky, Josh. 2018. *Reassembling Rubbish: Worlding Electronic Waste*. Cambridge, MA: MIT Press.

Lepawsky, Josh, Erin Araujo, John-Michael Davis, and Ramzy Kahhat. 2017. "Best of Two Worlds? Towards Ethical Electronics Repair, Reuse, Repurposing and Recycling." *Geoforum* 81 (May): 87–99. https://doi.org/10.1016/j.geoforum.2017.02.007.

Lerner, Steve. 2012. *Sacrifice Zones: The Front Lines of Toxic Chemical Exposure in the United States*. Cambridge, MA: MIT Press.

Levien, Michael. 2011. "Special Economic Zones and Accumulation by Dispossession in India." *Journal of Agrarian Change* 11 (4): 454–483. https://doi.org/10.1111/j.1471-0366.2011.00329.x.

Lewis, Tania. 2016. *Green Asia: Ecocultures, Sustainable Lifestyles, and Ethical Consumption*. Abingdon, UK: Routledge.

Li, Tania Murray. 2007. *The Will to Improve: Governmentality, Development, and the Practice of Politics*. Durham, NC: Duke University Press.

Liboiron, Max. 2013. "Modern Waste as Strategy." *Lo Squaderno: Explorations in Space and Society* 29:9–12. https://maxliboiron.files.wordpress.com/2013/08/liboiron-modern-waste-as-strategy-extracted1.pdf.

Liboiron, Max. 2021. *Pollution Is Colonialism*. Durham, NC: Duke University Press.

Liboiron, Max, and Josh Lepawsky. 2022. *Discard Studies: Wasting, Systems, and Power*. Cambridge, MA: MIT Press.

Loschi, Chiara. 2019. "Local Mobilisations and the Formation of Environmental Networks in a Democratizing Tunisia." *Social Movement Studies* 18 (1): 93–112. https://doi.org/10.1080/14742837.2018.1540974.

Luque-Ayala, Andrés, Simon Marvin, and H. Bulkeley. 2018. *Rethinking Urban Transitions*. Abingdon, UK: Routledge.

Luthra, Aman. 2015. "Modernity's Garbage: Struggles over Municipal Solid Waste in Urban India." PhD thesis, Johns Hopkins University, Baltimore, MD.

Luthra, Aman. 2018. "'Old Habits Die Hard': Discourses of Urban Filth in *Swachh Bharat Mission* and *The Ugly Indian*." *Journal of Multicultural Discourses* 13 (2): 120–138. https://doi.org/10.1080/17447143.2018.1467917.

Luthra, Aman. 2019. "Municipalization for Privatization's Sake." *Society and Business Review* 14 (2): 135–154. https://doi.org/10.1108/SBR-11-2017-0102.

Luthra, Aman. 2020. "Housewives and Maids: The Labor of Household Recycling in Urban India." *Environment and Planning E: Nature and Space* 4 (2): 475–498. https://doi.org/10.1177/2514848620914219.

Luthra, Aman, and William Monteith. 2021. "Of Market Vendors and Waste Collectors: Labour, Informality, and Aesthetics in the Era of World-Class City Making." *Antipode.* https://doi.org/10.1111/anti.12784.

Lutringer, Christine. 2017. "Environmental and Social Justice in Solid Waste Management." *International Development Policy | Revue Internationale de Politique de Développement* 8, no. 2 (September). https://doi.org/10.4000/poldev.2487.

Lutringer, Christine, and Shalini Randeria. 2017. "How Not to Waste a Garbage Crisis: Food Consumption, Solid Waste Management and Civic Activism in Bangalore/Bengaluru, India." *International Development Policy | Revue Internationale de Politique de Développement* 8, no. 2 (September). https://doi.org/10.4000/poldev.2476.

MacBride, Samantha. 2011. *Recycling Reconsidered: The Present Failure and Future Promise of Environmental Action in the United States.* Cambridge, MA: MIT Press.

MacGregor, Sherilyn. 2007. *Beyond Mothering Earth: Ecological Citizenship and the Politics of Care.* Vancouver: UBC Press.

MacGregor, Sherilyn. 2021a. "Making Matter Great Again? Ecofeminism, New Materialism and the Everyday Turn in Environmental Politics." *Environmental Politics* 30 (1/2): 41–60. https://doi.org/10.1080/09644016.2020.1846954.

MacGregor, Sherilyn. 2021b. "Finding Transformative Potential in the Cracks? The Ambiguities of Urban Environmental Activism in a Neoliberal City." *Social Movement Studies* 20 (3): 329–345. https://doi.org/10.1080/14742837.2019.1677224.

Mah, Alice. 2021. "Future-Proofing Capitalism: The Paradox of the Circular Economy for Plastics." *Global Environmental Politics* 21 (2): 121–142. https://doi.org/10.1162/glep_a_00594.

Mah, Alice. 2022. *Plastic Unlimited: How Corporations Are Fuelling the Ecological Crisis and What We Can Do About It.* Cambridge, UK: Polity.

Maldonado-Torres, Nelson. 2011. "Thinking through the Decolonial Turn: Post-Continental Interventions in Theory, Philosophy, and Critique—An Introduction." TRANSMODERNITY: *Journal of Peripheral Cultural Production of the Luso-Hispanic World* 1 (2). https://doi.org/10.5070/T412011805.

Malier, Hadrien. 2019. "Greening the Poor: The Trap of Moralization." *British Journal of Sociology* 70 (5). https://doi.org/10.1111/1468-4446.12672.

Maniates, Michael F. 2001. "Individualization: Plant a Tree, Buy a Bike, Save the World?" *Global Environmental Politics* 1 (3): 31–52. https://doi.org/doi:10.1162/152638001316881395.

Manzi, Maya, Joilson Santos Santana, and Cristina Maria Dacach Fernandez Marchi. 2022. "'Accumulation by Appropriation': The Integration of Recyclable-Waste Collector Cooperatives in Salvador, Brazil, and the Right to the City." *Environment and Planning D: Society and Space* 40 (4): 683–705. https://doi.org/10.1177/02637758221110882.

Martinez-Alier, Joan. 2003. *The Environmentalism of the Poor: A Study of Ecological Conflicts and Valuation.* Cheltenham, UK: Edward Elgar.

Martinez-Alier, Joan. 2016. "Global Environmental Justice and the Environmentalism of the Poor." In *The Oxford Handbook of Environmental Political Theory,* edited by Teena Gabrielson et al., 547–562. Oxford: Oxford University Press. https://doi.org/10.1093/oxfordhb/9780199685271.013.25.

Mathew, Melvin. 2022. "Dry Waste Collectors Overworked." *Bangalore Mirror,* April 28, 2022. https://bangaloremirror.indiatimes.com/bangalore/civic/dry-waste-collectors-overworked/articleshow/91133073.cms.

McClintock, Nathan. 2018. "Cultivating (a) Sustainability Capital: Urban Agriculture, Ecogentrification, and the Uneven Valorization of Social Reproduction." *Annals of the American Association of Geographers* 108 (2): 579–590. https://doi.org/10.1080/24694452.2017.1365582.

McFarlane, Colin. 2008. "Sanitation in Mumbai's Informal Settlements: State, 'Slum,' and Infrastructure." *Environment and Planning A* 40 (1): 88–107. https://doi.org/10.1068/a39221.

McGinty, David. 2020. "How to Build a Circular Economy." World Resources Institute, August 6, 2020. https://www.wri.org/insights/how-build-circular-economy.

McGranahan, Gordon, and Diana Mitlin. 2016. "Learning from Sustained Success: How Community-Driven Initiatives to Improve Urban Sanitation Can Meet the Challenges." *World Development* 87:307–317. https://doi.org/10.1016/j.worlddev.2016.06.019.

McLaren, Duncan, and Julian Agyeman. 2015. *Sharing Cities: A Case for Truly Smart and Sustainable Cities.* Cambridge, MA: MIT Press.

Meagher, Kate. 2013. "Unlocking the Informal Economy: A Literature Review on Linkages between Formal and Informal Economies in Developing Countries." WIEGO Working Paper 27. https://www.wiego.org/publications/literature-review-linkages-between-formal-and-informal-economies-developing-countries.

Melamed, Jodi. 2015. "Racial Capitalism." *Critical Ethnic Studies* 1 (1): 76–85. https://doi.org/10.5749/jcritethnstud.1.1.0076.

Melo, Carla. 2019. "Your Trash Is My Sustenance: Recycling the Image of 'Waste Pickers.'" In *Sustainable Tools for Precarious Times: Performance Actions in the Americas,* edited by Natalie Alvarez, Claudette Lauzon, and Keren Zaiontz, 151–175. Cham: Springer Nature Switzerland.

Melosi, Martin V. 2004. *Garbage in the Cities: Refuse Reform and the Environment*. Pittsburgh, PA: University of Pittsburgh Press.

Menezes, Naveen. 2022a. "BBMP Seeks Legal Sanction for New Waste Management Agency." *Deccan Herald*, July 4, 2022. https://www.deccanherald.com/city/top-bengaluru -stories/bbmp-seeks-legal-sanction-for-new-waste-management-agency-1123842.html.

Menezes, Naveen. 2022b. "Bengaluru's Waste Collection Vehicles to Turn Green." *Deccan Herald*, August 24, 2022. https://www.deccanherald.com/city/top-bengaluru -stories/bengaluru-s-waste-collection-vehicles-to-turn-green-1138731.html.

Menon, Harish C. 2017. "How Bangalore Went from Being India's Most Liveable City to a Dystopia in the Making." Text. Scroll.In. August 17, 2017. https://scroll.in/article /847397/how-bangalore-went-from-being-indias-most-liveable-city-to-a-dystopia-in -the-making.

Merchant, Carolyn. 2003. "Shades of Darkness: Race and Environmental History." *Environmental History* 8 (3): 380–394. https://doi.org/10.2307/3986200.

Mi, Zhifu, Jiali Zheng, Jing Meng, Heran Zheng, Xian Li, D'Maris Coffman, Johan Woltjer, Shouyang Wang, and Dabo Guan. 2019. "Carbon Emissions of Cities from a Consumption-Based Perspective." *Applied Energy* 235:509–518. https://doi.org/10 .1016/j.apenergy.2018.10.137.

Michael, Kavya, and Vamsi Vakulabharanam. 2016. "Class and Climate Change in Post-Reform India." *Climate and Development* 8 (3): 224–233. https://doi.org/10.1080 /17565529.2015.1034235.

Michaelson, Ruth. 2021. "'Waste Colonialism': World Grapples with West's Unwanted Plastic." *The Guardian*, December 31, 2021, sec. Environment. https://www.theguardian .com/environment/2021/dec/31/waste-colonialism-countries-grapple-with-wests -unwanted-plastic.

Mignolo, Walter D. 2007. "Delinking: The Rhetoric of Modernity, the Logic of Coloniality and the Grammar of de-Coloniality." *Cultural Studies* 21 (2/3): 449–514. https://doi.org/10.1080/09502380601162647.

Millar, Kathleen M. 2018. *Reclaiming the Discarded: Life and Labor on Rio's Garbage Dump*. Durham, NC: Duke University Press.

Millington, Nate, and Suraya Scheba. 2020. "Day Zero and the Infrastructures of Climate Change: Water Governance, Inequality, and Infrastructural Politics in Cape Town's Water Crisis." *International Journal of Urban and Regional Research* 45, no. 1 (March): 116–132. https://doi.org/10.1111/1468-2427.12899.

Millington, Nate, Kathleen Stokes, and Mary Lawhon. 2022. "Whose Value Lies in the Urban Mine? Reconfiguring Permissions, Work, and the Benefits of Waste in South Africa." *Annals of the American Association of Geographers* 112 (7): 1942–1957. https://doi.org/10.1080/24694452.2022.2042181.

MindTree. n.d. "I Got Garbage 2017–2018 | Mindtree." Mindtree.Com. Accessed October 16, 2022. https://www.mindtree.com/about/investors/annual-reports/annual -report-2017-2018/tree-always-gives-back/i-got-garbage.

Mindtree Ltd., dir. 2014. *I Got Garbage—a Mindtree Initiative*. YouTube video. https:// www.youtube.com/watch?v=LsaGE1A5wl8.

Miraftab, Faranak. 2004a. "Making Neo-Liberal Governance: The Disempowering Work of Empowerment." *International Planning Studies* 9 (4): 239–259. https://doi .org/10.1080/13563470500050130.

Miraftab, Faranak. 2004b. "Neoliberalism and Casualization of Public Sector Services: The Case of Waste Collection Services in Cape Town, South Africa." *International Journal of Urban and Regional Research* 28 (4): 874–892. https://doi.org/10.1111/j.0309 -1317.2004.00557.x.

Miraftab, Faranak, and Shana Wills. 2005. "Insurgency and Spaces of Active Citizenship: The Story of Western Cape Anti-Eviction Campaign in South Africa." *Journal of Planning Education and Research* 25 (2): 200–217. http://dx.doi.org/10.1177/0739456X05282182.

Mitlin, Diana. 2008. "With and beyond the State—Co-Production as a Route to Political Influence, Power and Transformation for Grassroots Organizations." *Environment and Urbanization* 20 (2): 339–360. https://doi.org/10.1177/0956247808096117.

Mitlin, Diana, and Sheridan Bartlett. 2018. "Editorial: Co-Production—Key Ideas." *Environment and Urbanization* 30 (2): 355–366. https://doi.org/10.1177/0956247818791931.

Moore, Sarah A. 2012. "Garbage Matters Concepts in New Geographies of Waste." *Progress in Human Geography* 36 (6): 780–799. https://doi.org/10.1177/0309132512437077.

Moretto, Luisa, Giuseppe Faldi, Marco Ranzato, Federica Natalia Rosati, Jean-Pierre Ilito Boozi, and Jacques Teller. 2018. "Challenges of Water and Sanitation Service Co-Production in the Global South." *Environment and Urbanization* 30 (2): 425–443. https://doi.org/10.1177/0956247818790652.

Mudliar, Pranietha, and Tomas Koontz. 2018. "The Muting and Unmuting of Caste across Inter-Linked Action Arenas: Inequality and Collective Action in a Community-Based Watershed Group." *International Journal of the Commons* 12 (1): 225–248. http://doi.org/10.18352/ijc.807.

Murphy, Joseph, and Sarah Parry. 2021. "Gender, Households and Sustainability: Disentangling and Re-Entangling with the Help of 'Work' and 'Care.'" *Environment and Planning E: Nature and Space* 4, no. 3 (January): 1099–1120. https://doi.org/10 .1177/2514848620948432.

Nagendra, Harini. 2016. *Nature in the City: Bengaluru in the Past, Present, and Future*. Oxford: Oxford University Press.

Nagendra, Harini, and Divya Gopal. 2010. "Street Trees in Bangalore: Density, Diversity, Composition and Distribution." *Urban Forestry and Urban Greening* (special

section, "Forest Recreation and Nature Tourism") 9 (2): 129–137. https://doi.org/10
.1016/j.ufug.2009.12.005.

Nagendra, Harini, and Elinor Ostrom. 2014. "Applying the Social-Ecological System
Framework to the Diagnosis of Urban Lake Commons in Bangalore, India." *Ecology
and Society* 19 (2). https://www.jstor.org/stable/26269569.

Nair, Janaki. 2005. *The Promise of the Metropolis: Bangalore's Twentieth Century.* Oxford:
Oxford University Press.

Narayan, Priti. 2015. "Patterns in Arbitrariness: Resettlement Experiences of the
Unrecognized Urban Poor in Chennai." In *Development-Induced Displacement and
Resettlement,* edited by Irge Satiroglu and Narae Choi, 170–183. London: Routledge.

Narayanareddy, Rajyashree. 2011. "Specters of Waste in India's 'Silicon Valley': The
Underside of Bangalore's Hi-Tech Economy." PhD Thesis. Minneapolis: University
of Minnesota.

Newell, Peter, and Matthew Paterson. 2010. *Climate Capitalism: Global Warming and
the Transformation of the Global Economy.* Cambridge, UK: Cambridge University Press.

News Minute, The. 2021. "Civic Workers in Bengaluru Demand Protective Equipment
amid Lockdown." April 28, 2021. https://www.thenewsminute.com/article/civic
-workers-bengaluru-demand-protective-equipment-amid-lockdown-147997.

Nielsen, Kristian S., Kimberly A. Nicholas, Felix Creutzig, Thomas Dietz, and Paul C.
Stern. 2021. "The Role of High-Socioeconomic-Status People in Locking In or Rapidly Reducing Energy-Driven Greenhouse Gas Emissions." *Nature Energy* 6 (11):
1011–1116. https://doi.org/10.1038/s41560-021-00900-y.

Nirupama, V. 2016. "Collection of Dry Waste: Non-Profit Organisations Cry Unfair
Competition from Big Corporate—The Economic Times." *Economic Times,* December 12,
2016. https://economictimes.indiatimes.com/news/politics-and-nation/collection
-of-dry-waste-non-profit-organisations-cry-unfair-competition-from-big-corporate
/articleshow/55934994.cms.

Nisbett, Nicholas. 2020. *Growing Up in the Knowledge Society: Living the IT Dream in
Bangalore.* New Delhi: Routledge.

Nixon, Rob. 2011. *Slow Violence and the Environmentalism of the Poor.* Cambridge,
MA: Harvard University Press.

OECD. 2007. "Municipal Waste Generation—Outlook from OECD." Indicator Assessment. 2007. http://www.eea.europa.eu/data-and-maps/indicators/municipal-waste
-generation-outlook-from-oecd/municipal-waste-generation-outlook-from.

O'Hare, Patrick. 2020. "'We Looked After People Better When We Were Informal':
The 'Quasi-Formalisation' of Montevideo's Waste-Pickers." *Bulletin of Latin American
Research* 39 (1): 53–68. https://doi.org/10.1111/blar.12957.

O'Hare, Patrick. 2022. *Rubbish Belongs to the Poor: Hygienic Enclosure and the Waste Commons*. 1st ed. London: Pluto Press.

O'Neill, Kate. 2019. *Waste*. Cambridge, UK: Polity.

Ostrom, Elinor. 1990. *Governing the Commons: The Evolution of Institutions for Collective Action*. Cambridge, UK: Cambridge University Press.

Ostrom, Elinor. 1993. "A Communitarian Approach to Local Governance." *National Civic Review* 82 (3): 226–233. https://doi.org/10.1002/ncr.4100820305.

Ostrom, Elinor. 1996. "Crossing the Great Divide: Coproduction, Synergy, and Development." *World Development* 24 (6): 1073–1087. https://doi.org/10.1016/0305 -750X(96)00023-X.

Oxfam International. 2019. "India Extreme Inequality in Numbers." Oxfam International. October 19, 2019. https://www.oxfam.org/en/india-extreme-inequality-numbers.

Pande, Raksha. 2016. "Strategic Essentialism." In *International Encyclopedia of Geography: People, the Earth, Environment and Technology*. New Jersey, USA: Wiley and the American Association of Geographers.

Pani, Narendar, Sindhu Radhakrishna, and Kishor G. Bhat, eds. 2010. *Bengaluru, Bangalore, Bengaluru: Imaginations and Their Times*. Los Angeles: Sage.

Park, Lisa, and David N. Pellow. 2011. *The Slums of Aspen*. New York: New York University Press.

Parnell, Susan, and Jennifer Robinson. 2012. "(Re)Theorizing Cities from the Global South: Looking beyond Neoliberalism." *Urban Geography* 33 (4): 593–617. https://doi .org/10.2747/0272-3638.33.4.593.

Patel, Raj, and Jason W Moore. 2018. *A History of the World in Seven Cheap Things: A Guide to Capitalism, Nature, and the Future of the Planet*. Berkeley: University of California Press.

Pathak, Gauri. 2020. "'Plastic Pollution' and Plastics as Pollution in Mumbai, India." *Ethnos* October 28, 2020, pp. 1–20. https://doi.org/10.1080/00141844.2020.1839116.

Pathak, Gauri. 2021. "Nation Branding, Soft Hindutva, and Ecotraditionalism in Anti-Plastics Discourses in India." *Identities* 29 (6):768–786. https://doi.org/10.1080 /1070289X.2021.1920773.

Pathak, Gauri, and Mark Nichter. 2021. "Ecocommunicability, Citizenship, and Discourses on Plastic Control in India." *Geoforum* 125 (October): 132–139. https://doi .org/10.1016/j.geoforum.2021.04.027.

Paulson, Susan, Giacomo D'Alisa, Federico Demaria, and Giorgos Kallis. 2020. *The Case for Degrowth*. Cambridge, UK: Polity.

Peck, Jamie. 2010. *Constructions of Neoliberal Reason*. Oxford: Oxford University Press.

Peck, Jamie. 2012. "Austerity Urbanism." *City* 16 (6): 626–655. https://doi.org/10.1080
/13604813.2012.734071.

Pellow, David N. 2004. *Garbage Wars: The Struggle for Environmental Justice in Chicago.*
Cambridge, MA: MIT Press.

Pellow, David N. 2016. "Toward a Critical Environmental Justice Studies: Black
Lives Matter as an Environmental Justice Challenge." *Du Bois Review: Social Science
Research on Race* 13 (2): 221–236. https://doi.org/10.1017/S1742058X1600014X.

Pieterse, Edgar. 2008. *City Futures: Confronting the Crisis of Urban Development.* London:
Bloomsbury.

Pieterse, Edgar. 2021. "Unlocking Political Potentialities." In *Global Urbanism: Knowledge, Power, and the City*, edited by Michele Lancione and Colin McFarlane, 193–202.
London: Routledge.

Porras Bulla, Julián, Michael Rendon, and Josep Espluga Trenc. 2021. "Policing the
Stigma in Our Waste: What We Know about Informal Waste Pickers in the Global
North." *Local Environment* 26 (10): 1299–1312. https://doi.org/10.1080/13549839
.2021.1974368.

Portney, Kent E. 2013. *Taking Sustainable Cities Seriously: Economic Development, the Environment, and Quality of Life in American Cities.* Cambridge, MA: MIT Press.

Prashad, Vijay. 2001. "The Technology of Sanitation in Colonial Delhi." *Modern Asian
Studies* 35 (1): 113–155. https://doi.org/10.1017/S0026749X01003626.

Prasher, Garima. 2022. "Wafted Away." *Bangalore Mirror*, October 12, 2022. https://
bangaloremirror.indiatimes.com/bangalore/civic/wafted-away/articleshow/94795522
.cms.

Pulido, Laura. 1996. "Ecological Legitimacy and Cultural Essentialism: Hispano Grazing in the South West." *Capitalism Nature Socialism* 7 (4): 37–58. https://doi.org/10
.1080/10455759609358707.

Pulido, Laura. 2015. "Geographies of Race and Ethnicity 1: White Supremacy vs White
Privilege in Environmental Racism Research." *Progress in Human Geography* 39 (6):
809–817. https://doi.org/10.1177/0309132514563008.

Pulido, Laura. 2017. "Geographies of Race and Ethnicity II: Environmental Racism,
Racial Capitalism and State-Sanctioned Violence." *Progress in Human Geography* 41 (4):
524–533. https://doi.org/10.1177/0309132516646495.

Purkayastha, Debapratim, and Rajiv Fernando. 2007. "The Body Shop: Social Responsibility or Sustained Greenwashing?" In *Case Studies in Sustainability Management and
Strategy*, edited by Jost Hamschmidt. London: Routledge.

Rademacher, Anne M. 2011. *Reigning the River.* Durham, NC: Duke University
Press.

Rademacher, Anne M., and K. Sivaramakrishnan. 2013. *Ecologies of Urbanism in India: Metropolitan Civility and Sustainability*. Hong Kong: Hong Kong University Press.

Rajamani, Lavanya. 2007. "Public Interest Environmental Litigation in India: Exploring Issues of Access, Participation, Equity, Effectiveness and Sustainability." *Journal of Environmental Law* 19 (3): 293–321. https://doi.org/10.1093/jel/eqm020.

Rajan, M. C. 2019. "MBAs, Engineers among 4,000 Applicants Competing for 14 Sweeper Jobs in TN." *Hindustan Times*, February 5, 2019. https://www.hindustan times.com/education/over-4000-applicants-including-mtech-and-mba-apply -for-14-sanitary-workers-posts-in-tamil-nadu/story-vZSe3csNzzx5h3ZnsHk3gO .html.

Ram, Theja. 2019. "Bengaluru's Garbage Doubled in Last Five Years, City Churns Almost 6000 Tonnes per Day | The News Minute." *News Minute*, May 18, 2019. https:// www.thenewsminute.com/article/bengalurus-garbage-doubled-last-five-years-city -churns-almost-6000-tonnes-day-102564.

Ramakrishnan, Anjali, Matthias Kalkuhl, Sohail Ahmad, and Felix Creutzig. 2020. "Keeping Up with the Patels: Conspicuous Consumption Drives the Adoption of Cars and Appliances in India." *Energy Research and Social Science* 70 (December): 101742. https://doi.org/10.1016/j.erss.2020.101742.

Ramani, Chitra V. 2012. "Mavallipura Residents Up Their Ante." *Hindu*, July 18, 2012. http://www.thehindu.com/news/cities/bangalore/article3652334.ece.

Ramani, Chitra V. 2017. "Once Waste Pickers, Now Successful Managers." *Hindu*, December 27, 2017, sec. Bengaluru. https://www.thehindu.com/news/cities/bangalore /once-waste-pickers-now-successful-managers/article22289632.ece.

Ramaswami, Anu. 2020. "Unpacking the Urban Infrastructure Nexus with Environment, Health, Livability, Well-Being, and Equity." *One Earth* 2 (2): 120–124. https:// doi.org/10.1016/j.oneear.2020.02.003.

Ranganathan, Malini. 2014. "Paying for Pipes, Claiming Citizenship: Political Agency and Water Reforms at the Urban Periphery." *International Journal of Urban and Regional Research* 38 (2): 590–608. https://doi.org/10.1111/1468-2427.12028.

Ranganathan, Malini. 2016. "Thinking with Flint: Racial Liberalism and the Roots of an American Water Tragedy." *Capitalism Nature Socialism* 27 (3): 17–33. https://doi .org/10.1080/10455752.2016.1206583.

Ranganathan, Malini. 2021. "Caste, Racialization, and the Making of Environmental Unfreedoms in Urban India." *Ethnic and Racial Studies* 45 (2): 257–277. https://doi .org/10.1080/01419870.2021.1933121.

Ranganathan, Malini. 2022. "Analysis | Urban Flooding Has Everything to Do with Real Estate Corruption." *Washington Post*, September 9, 2022. https://www .washingtonpost.com/politics/2022/09/09/flooding-pakistan-india-development.

Ray, Raka, and Seemin Qayuum. 2009. *Cultures of Servitude: Modernity, Domesticity, and Class in India*. Stanford, CA: Stanford University Press.

Reckwitz, Andreas. 2002. "Toward a Theory of Social Practices: A Development in Culturalist Theorizing." *European Journal of Social Theory* 5 (2): 243–263. https://doi.org/10.1177/13684310222225432.

Reddy, Rajyashree. 2015. "Producing Abjection: E-Waste Improvement Schemes and Informal Recyclers of Bangalore." *Geoforum* 62 (June): 166–174. https://doi.org/10.1016/j.geoforum.2015.04.003.

Reddy, Rajyashree. 2016. "Reimagining E-Waste Circuits: Calculation, Mobile Policies, and the Move to Urban Mining in Global South Cities." *Urban Geography* 37 (1): 57–76. https://doi.org/10.1080/02723638.2015.1046710.

Reddy, Rajyashree. 2021. "Of Holy Cows and Unholy Politics: Dalits, Annihilation and More-than-Human Urban Abolition Ecologies." *International Journal of Urban and Regional Research* 45 (4): 643–657. https://doi.org/10.1111/1468-2427.13028.

Rice, Jennifer L., Daniel Aldana Cohen, Joshua Long, and Jason R. Jurjevich. 2020. "Contradictions of the Climate-Friendly City: New Perspectives on Eco-Gentrification and Housing Justice." *International Journal of Urban and Regional Research* 44 (1): 145–165. https://doi.org/10.1111/1468-2427.12740.

Rigon, Andrea, and Vanesa Castán Broto. 2021. *Inclusive Urban Development in the Global South: Intersectionality, Inequalities, and Community*. Abingdon, UK: Routledge.

Robinson, Cedric J. 2000. *Black Marxism: The Making of the Black Radical Tradition*. Chapel Hill: University of North Carolina Press.

Rodrigues, Valerian. 2009. "Untouchability, Filth, and the Public Domain." In *Humiliation: Claims and Context*, edited by Gopal Guru, 108–123. New Delhi: Oxford University Press.

Rosaldo, Manuel. 2016. "Revolution in the Garbage Dump: The Political and Economic Foundations of the Colombian Recycler Movement, 1986–2011." *Social Problems* 63 (3): 351–372. https://doi.org/10.1093/socpro/spw015.

Rosaldo, Manuel. 2019. "The Antinomies of Successful Mobilization: Colombian Recyclers Manoeuvre between Dispossession and Exploitation." *Development and Change*, July. https://onlinelibrary.wiley.com/doi/abs/10.1111/dech.12536.

Rosaldo, Manuel, and La Vida Alegre. 2016. "Pimp My Carroça Bogotá." *Berkeley Journal of Sociology*, September 27, 2016. http://berkeleyjournal.org/2016/09/pimp-my-carroca-bogota.

Rosan, Christina D., and Hamil Pearsall. 2018. *Growing a Sustainable City?: The Question of Urban Agriculture*. Toronto: University of Toronto Press.

Roy, Ananya. 2009a. "Civic Governmentality: The Politics of Inclusion in Beirut and Mumbai." *Antipode* 41 (1): 159–179. https://doi.org/10.1111/j.1467-8330.2008.00660.x.

Roy, Ananya. 2009b. "Why India Cannot Plan Its Cities: Informality, Insurgence and the Idiom of Urbanization." *Planning Theory* 8 (1): 76–87. https://doi.org/10.1177/1473095208099299.

Roy, Ananya. 2017. "The Infrastructure of Assent: Professions in the Age of Trumpism." *Avery Review*, January 20, 2017. https://averyreview.com/issues/21/the-infrastructure-of-assent.

Roy, Ananya, and Aihwa Ong, eds. 2011. *Worlding Cities: Asian Experiments and the Art of Being Global*. Chichester, UK: Wiley-Blackwell.

Rufo, Christopher F. 2019. "Emerald Trash Heap." *City Journal*, May 13, 2019. https://www.city-journal.org/seattle-trash-crisis.

Sahakian, Marlyne, and Béatrice Bertho. 2018. "Exploring Emotions and Norms around Swiss Household Energy Usage: When Methods Inform Understandings of the Social." *Energy Research and Social Science* 45:81–90. https://doi.org/10.1016/j.erss.2018.06.017.

Sahakian, Marlyne, and Harold Wilhite. 2014. "Making Practice Theory Practicable: Towards More Sustainable Forms of Consumption." *Journal of Consumer Culture* 14 (1): 25–44. https://doi.org/10.1177/1469540513505607.

Salve, Pradeep S., Dhananjay W. Bansod, and Hemangi Kadlak. 2017. "Safai Karamcharis in a Vicious Cycle." *Economic and Political Weekly* 52 (13): 37. https://www.epw.in/journal/2017/13/perspectives/safai-karamcharis-avicious-cycle.html.

Samson, Melanie. 2009. "Wasted Citizenship? Reclaimers and the Privatised Expansion of the Public Sphere." *Africa Development* 34 (3/4): 1–25. https://www.ajol.info/index.php/ad/article/view/63525.

Samson, Melanie. 2010. "Reclaiming Reusable and Recyclable Materials in Africa." WIEGO Working Paper 16. https://www.wiego.org/publications/reclaiming-reusable-and-recyclable-materials-africa-critical-review-english-language-li.

Samson, Melanie. 2015. "Accumulation by Dispossession and the Informal Economy—Struggles over Knowledge, Being and Waste at a Soweto Garbage Dump." *Environment and Planning D: Society and Space* 33 (5): 813–830. https://doi.org/10.1177/0263775815600058.

Samson, Melanie. 2016. "Old Trash, New Ideas: Public Waste Management and Informal Reclaimers." In *Making Public in a Privatised World: The Struggle for Essential Services*, edited by David A. McDonald, 41–58. Camden, UK: Bloomsbury.

Samson, Melanie. 2019. "Whose Frontier Is It Anyway? Reclaimer 'Integration' and the Battle over Johannesburg's Waste-Based Commodity Frontier." *Capitalism Nature Socialism* 31 (December): 60–75. https://doi.org/10.1080/10455752.2019.1700538.

Savini, Federico. 2019. "The Economy That Runs on Waste: Accumulation in the Circular City." *Journal of Environmental Policy and Planning* 21 (6): 675–691. https://doi.org/10.1080/1523908X.2019.1670048.

Saxenian, AnnaLee. 2002. "Silicon Valley's New Immigrant High-Growth Entrepreneurs." *Economic Development Quarterly* 16 (1): 20–31. https://doi.org/10.1177/0891242402016001003.

Saxenian, AnnaLee. 2005. "From Brain Drain to Brain Circulation: Transnational Communities and Regional Upgrading in India and China." *Studies in Comparative International Development* 40 (2): 35–61. https://doi.org/10.1007/BF02686293.

Schatzki, Theodore R. 1996. *Social Practices: A Wittgensteinian Approach to Human Activity and the Social.* Cambridge, UK: Cambridge University Press.

Schenck, Rinie, and Phillip Frederick Blaauw. 2011. "The Work and Lives of Street Waste Pickers in Pretoria—a Case Study of Recycling in South Africa's Urban Informal Economy." *Urban Forum* 22:411–430. https://doi.org/10.1007/s12132-011-9125-x.

Schindler, Seth, and Federico Demaria. 2019. "'Garbage Is Gold': Waste-Based Commodity Frontiers, Modes of Valorization and Ecological Distribution Conflicts." *Capitalism Nature Socialism* 31 (4): 52–59. https://doi.org/10.1080/10455752.2019.1694553.

Schindler, Seth, Federico Demaria, and Shashi B. Pandit. 2012. "Delhi's Waste Conflict." *Economic and Political Weekly* 47 (42): 18–21. https://www.jstor.org/stable/41720261.

Schlosberg, David, and Luke Craven. 2019. *Sustainable Materialism: Environmental Movements and the Politics of Everyday Life.* Oxford: Oxford University Press.

Schmidt, Tobias S. 2014. "Low-Carbon Investment Risks and De-Risking." *Nature Climate Change* 4 (4): 237–239. https://doi.org/10.1038/nclimate2112.

Schroeder, Patrick, Kartika Anggraeni, and Uwe Weber. 2019. "The Relevance of Circular Economy Practices to the Sustainable Development Goals." *Journal of Industrial Ecology* 23 (1): 77–95. https://doi.org/10.1111/jiec.12732.

Schuller, Mark. 2009. "Gluing Globalization: NGOs as Intermediaries in Haiti." *PoLAR: Political and Legal Anthropology Review* 32 (1): 84–104. https://www.jstor.org/stable/24497527.

Schultz, Susanne. 2006. "Dissolved Boundaries and 'Affective Labor': On the Disappearance of Reproductive Labor and Feminist Critique in Empire." *Capitalism Nature Socialism* 17 (1): 77–82. https://doi.org/10.1080/10455750500505473.

Schwenkel, Christina. 2015. "Spectacular Infrastructure and Its Breakdown in Socialist Vietnam." *American Ethnologist* 42 (3): 520–534. https://doi.org/10.1111/amet.12145.

Scott, James C. 1985. *Weapons of the Weak: Everyday Forms of Peasant Resistance.* New Haven, CT: Yale University Press.

Sen, Amrita, and Harini Nagendra. 2020. "Local Community Engagement, Environmental Placemaking and Stewardship by Migrants: A Case Study of Lake Conservation in Bengaluru, India." *Landscape and Urban Planning* 204 (December): 103933. https://doi.org/10.1016/j.landurbplan.2020.103933.

Sen, Amrita, Hita Unnikrishnan, and Harini Nagendra. 2021. "Restoration of Urban Water Commons: Navigating Social-Ecological Fault Lines and Inequities." *Ecological Restoration* 39 (1/2): 120–129. https://muse.jhu.edu/article/793665.

Sensing Local. 2022. "Re-Visioning Bengaluru's DWCCs, by Sensing Local—ArchitectureLive!" February 22, 2022. https://architecture.live/re-visioning-bengalurus-dwccs-by-sensing-local.

Seyfang, Gill, and Alex Haxeltine. 2012. "Growing Grassroots Innovations: Exploring the Role of Community-Based Initiatives in Governing Sustainable Energy Transitions." *Environment and Planning C: Government and Policy* 30 (3): 381–400. https://doi.org/10.1068/c10222.

Sonal, Shruti. 2023. "Cash from Trash: Women Waste Pickers Are Upcycling Their Lives." *Times of India*. January 31, 2023. https://timesofindia.indiatimes.com/india/cash-from-trash-women-waste-pickers-are-upcycling-their-lives/articleshow/97464752.cms.

Shankar, Shiva, and Kanthi Swaroop. 2021. "Manual Scavenging in India." *CASTE: A Global Journal on Social Exclusion* 2 (1): 67–76. https://doi.org/10.26812/caste.v2i1.299.

Sharma, Mukul. 2017a. "Brahmanical Activism as Eco-Casteism: Reading the Life Narratives of Bindeshwar Pathak, Sulabh International, and 'Liberated' Dalits." *Biography* 40, no. 1: 199–221. http://dx.doi.org/10.1353/bio.2017.0009.

Sharma, Mukul. 2017b. *Caste and Nature: Dalits and Indian Environmental Policies*. New Delhi: Oxford University Press.

Sharma, Sneha. 2022. *Waste(d) Collectors: Politics of Urban Exclusion in Mumbai*. New Rockford, ND: Transcript Publishing.

Shekhar, Divya. 2018. "Bengaluru Still Not a 'Plastic Free' City." *Economic Times*, June 1, 2018. https://economictimes.indiatimes.com/news/politics-and-nation/bengaluru-still-not-a-plastic-free-city/articleshow/64410546.cms?.

Shove, Elizabeth, Mika Pantzar, and Matt Watson. 2012. *The Dynamics of Social Practice: Everyday Life and How It Changes*. London: Sage.

Shree, D. N. 2017. "The Herculean Task of Reining in Bengaluru's Garbage Mafia." *Citizen Matters, Bengaluru* (blog). September 26, 2017. https://bengaluru.citizenmatters.in/bangalore-garbage-mafia-problems-solutions-21577.

Sibley, David. 1995. *Geographies of Exclusion: Society and Difference in the West*. London: Routledge.

Silva, Ana L. Patrício, Joana C. Prata, Tony R. Walker, Armando C. Duarte, Wei Ouyang, Damià Barceló, and Teresa Rocha-Santos. 2021. "Increased Plastic Pollution due to COVID-19 Pandemic: Challenges and Recommendations." *Chemical Engineering Journal* 405:126683. https://doi.org/10.1016/j.cej.2020.126683.

Silver, Jonathan. 2014. "Incremental Infrastructures: Material Improvisation and Social Collaboration across Post-Colonial Accra." *Urban Geography* 35 (6): 788–804. https://doi.org/10.1080/02723638.2014.933605.

Simone, A. M. 2004. "People as Infrastructure: Intersecting Fragments in Johannesburg." *Public Culture* 16 (3): 407–429. https://muse.jhu.edu/article/173743.

Slavin, Terry. 2019. "From Untouchables to Plastic Waste Entrepreneurs." Reuters Events | Sustainable Business. August 6, 2019. https://www.reutersevents.com/sustainability/untouchables-plastic-waste-entrepreneurs.

Smith, S. E. 2020. "Performative Environmentalism Won't Address Climate Change." *Bitch Media*. January 2020. https://www.bitchmedia.org/article/personal-will-not-save-you-environmentalism.

Sovacool, Benjamin K., and Alexander Dunlap. 2022. "Anarchy, War, or Revolt? Radical Perspectives for Climate Protection, Insurgency and Civil Disobedience in a Low-Carbon Era." *Energy Research and Social Science* 86 (April): 102416. https://doi.org/10.1016/j.erss.2021.102416.

Spivak, Gayatri Chakravorty. 2010. "Can the Subaltern Speak?" In *Can the Subaltern Speak? Reflections on the History of an Idea*, edited by Rosalind Morris, 21–78. New York: Columbia University Press.

Spivak, Gayatri Chakravorty. 1999. *A Critique of Postcolonial Reason*. Cambridge, MA: Harvard University Press.

Spivak, Gayatri Chakravorty. 1988. "Subaltern Studies: Deconstructing Historiography." In *Selected Subaltern Studies*, edited by Gayatri Chakravorty Spivak, and Ranajit Guha. 3–32. New York: Oxford University Press.

Sreenath, Shreyas. 2019. "Numbing Machines." *Economic and Political Weekly* 54 (47). https://epw.in/node/155806/pdf.

Srinivas, Tulasi. 2002. "Flush with Success: Bathing, Defecation, Worship, and Social Change in South India." *Space and Culture* 5 (4): 368–386. https://doi.org/10.1177/1206331202005004004.

Stahel, Walter R. 2016. "The Circular Economy." *Nature News* 531 (7595): 435. https://doi.org/10.1038/531435a.

Stamatopoulou-Robbins, Sophia. 2014. "Occupational Hazards." *Comparative Studies of South Asia, Africa and the Middle East* 34 (3): 476–496. https://muse.jhu.edu/article/566147.

Stamatopoulou-Robbins, Sophia. 2019. *Waste Siege: The Life of Infrastructure in Palestine*. Stanford, CA: Stanford University Press.

Stehlin, John G. 2019. *Cyclescapes of the Unequal City: Bicycle Infrastructure and Uneven Development*. Minneapolis: University of Minnesota Press.

Steinberg, Paul F., and Stacy D. VanDeveer. 2012. *Comparative Environmental Politics: Theory, Practice, and Prospects*. Cambridge, MA: MIT Press.

Stephens, Jennie C. 2020. *Diversifying Power: Why We Need Antiracist, Feminist Leadership on Climate and Energy*. Washington, DC: Island Press.

Sultana, Farhana. 2018. "Gender and Water in a Changing Climate: Challenges and Opportunities." In *Water Security across the Gender Divide*, edited by Christiane Frölich, Giovanna Gioli, Roger Cremades, and Henri Myrttinen, 17–33. Cham: Springer Nature.

Swaroop, Kanthi, and Joel Lee. 2021. "Caste and COVID-19." *Economic and Political Weekly* 56 (13): 35.

Swyngedouw, Erik. 2009. "The Antinomies of the Postpolitical City: In Search of a Democratic Politics of Environmental Production." *International Journal of Urban and Regional Research* 33 (3): 601–620. https://doi.org/10.1111/j.1468-2427.2009.00859.x.

Sze, Julie, ed. 2018. *Sustainability: Approaches to Environmental Justice and Social Power*. New York: New York University Press.

Talbott, Taylor Cass. 2019. "A Green Army Is Ready to Keep Plastic Waste Out of the Ocean." *Scientific American Blog*. October 7, 2019. https://blogs.scientificamerican.com/observations/a-green-army-is-ready-to-keep-plastic-waste-out-of-the-ocean.

Taylor Aiken, Gerald, Lucie Middlemiss, Susannah Sallu, and Richard Hauxwell-Baldwin. 2017. "Researching Climate Change and Community in Neoliberal Contexts: An Emerging Critical Approach." *Wiley Interdisciplinary Reviews: Climate Change* 8 (4): e463. https://doi.org/10.1002/wcc.463.

Teltumbde, Anand. 2015. "No Swachh Bharat without Annihilation of Caste." *Economic and Political Weekly* 49 (45): 7–8. https://www.epw.in/journal/2014/45/margin-speak/no-swachh-bharat-without-annihilation-caste.html.

Theodore, Nik. 2020. "Governing through Austerity: (Il)Logics of Neoliberal Urbanism after the Global Financial Crisis." *Journal of Urban Affairs* 42 (1): 1–17. https://doi.org/10.1080/07352166.2019.1623683.

Thompson, Andrea. 2019. "Waste Pickers in India Have a Brighter Future Thanks to the Body Shop." *Marie Claire*, May 9, 2019. https://www.marieclaire.co.uk/reports/waste-pickers-652770.

Times of India. 2014. "Mandur Blocks Roads, 200 Trash Trucks Return." June 16, 2014. http://timesofindia.indiatimes.com/city/bengaluru/Mandur-blocks-roads-200-trash-trucks-return/articleshow/36630985.cms.

Truelove, Yaffa. 2019. "Rethinking Water Insecurity, Inequality and Infrastructure through an Embodied Urban Political Ecology." *Wiley Interdisciplinary Reviews: Water* 6 (3): e1342. https://doi.org/10.1002/wat2.1342.

Truelove, Yaffa, and Emma Mawdsley. 2011. "Discourses of Citizenship and Criminality in Clean, Green Delhi." *A Companion to the Anthropology of India*, edited by Isabelle Clark-Decès, 407–425. Chichester, UK: Wiley-Blackwell.

Tsing, Anna Lowenhaupt. 2011. *Friction: An Ethnography of Global Connection*. Princeton, NJ: Princeton University Press.

Tuck, Eve. 2009. "Suspending Damage: A Letter to Communities." *Harvard Educational Review* 79 (3): 409–428. http://dx.doi.org/10.17763/haer.79.3.n0016675661t3n15.

Tucker, Jennifer Lee, and Manisha Anantharaman. 2020. "Informal Work and Sustainable Cities: From Formalization to Reparation." *One Earth* 3 (3): 290–299. https://doi .org/10.1016/j.oneear.2020.08.012.

Tucker, Jennifer Lee, and Ryan Thomas Devlin. 2019. "Uncertainty and the Governance of Street Vending: A Critical Comparison across the North/South Divide." *International Journal of Urban and Regional Research* 43 (3): 460–475. https://doi.org /10.1111/1468-2427.12792.

UN Environment. 2020. "Emissions Gap Report 2020." http://www.unep.org/emiss ions-gap-report-2020.

UN Environment. 2018. "Empowering Waste Workers for a Cleaner, Safer City." http://www.unep.org/news-and-stories/story/empowering-waste-workers-cleaner -safer-city.

Unnikrishnan, Hita, Vanesa Castán Broto, and Harini Nagendra. 2020. "Commons: Producing Collaborative Sustainable Urban Development: Experiences of Water Management in Bangalore, India." In *Dilemmas of Sustainable Urban Development: A View from Practice*, edited by Jonathan Metzger and Jenny Lindblad, 201–216. New York: Routledge.

Upadhya, Carol. 2008. "Rewriting the Code: Software Professionals and the Reconstitution of the Indian Middle Class Identity." In *Patterns of Middle Class Consumption in India and China*, edited by C. Jaffrelot and P. van der Veer. 55–87. New Delhi: Sage.

Upadhya, Carol. 2017. "Afterword: Middle Class Activism and Bangalore's Environmental Predicament." *International Development Policy | Revue Internationale de Politique de Développement* 8, no. 2 (September). https://doi.org/10.4000/poldev.2488.

Upadhya, Carol, and A. R. Vasavi, eds. 2008. *In an Outpost of the Global Economy: Work and Workers in India's Information Technology Industry*. New Delhi: Routledge.

Valette, Jim. 2021. "The New Coal: Plastics & Climate Change." Beyond Plastics at Bennington College. https://www.beyondplastics.org/plastics-and-climate.

Van Holstein, Ellen 2019. "The Social Boundary Work of New Middle-Class Organic Gardeners in Bangalore, India." *Urban Forestry and Urban Greening* 44 (August): 126432. https://doi.org/10.1016/j.ufug.2019.126432.

Vergara, Sintana E., and George Tchobanoglous. 2012. "Municipal Solid Waste and the Environment: A Global Perspective." *Annual Review of Environment and Resources* 37 (1): 277–309. https://doi.org/10.1146/annurev-environ-050511-122532.

Véron, René. 2006. "Remaking Urban Environments: The Political Ecology of Air Pollution in Delhi." *Environment and Planning A* 38 (11): 2093–2109. https://doi.org/10.1068/a37449.

Wachsmuth, David, and Hillary Angelo. 2018. "Green and Gray: New Ideologies of Nature in Urban Sustainability Policy." *Annals of the American Association of Geographers* 108 (4): 1038–1056. https://doi.org/10.1080/24694452.2017.1417819.

Wacquant, Loïc, Tom Slater, and Virgílio Borges Pereira. 2014. "Territorial Stigmatization in Action." *Environment and Planning A: Economy and Space* 46 (6): 1270–1280. https://doi.org/10.1068/a4606ge.

Walker, Gordon. 2011. "The Role for 'Community' in Carbon Governance." *Wiley Interdisciplinary Reviews: Climate Change* 2 (5): 777–782. https://doi.org/10.1002/wcc.137.

Warde, Alan. 2005. "Consumption and Theories of Practice." *Journal of Consumer Culture* 5 (2): 131–153. https://doi.org/10.1177/1469540505053090.

Watson, Vanessa. 2014. "Co-Production and Collaboration in Planning—The Difference." *Planning Theory and Practice* 15 (1): 62–76. https://doi.org/10.1080/14649357.2013.866266.

West, Cornel, and bell hooks. 2016. *Breaking Bread: Insurgent Black Intellectual Life.* New York: Routledge.

Wheeler, Kathryn, and Miriam Glucksmann. 2015. *Household Recycling and Consumption Work.* London: Palgrave Macmillan.

White, Damian F. 2019. "Ecological Democracy, Just Transitions and a Political Ecology of Design." *Environmental Values* 28 (1): 31–53. https://doi.org/10.3197/096327119X15445433913569.

Wiedmann, Thomas, Manfred Lenzen, Lorenz T. Keyßer, and Julia K. Steinberger. 2020. "Scientists' Warning on Affluence." *Nature Communications* 11 (1): 3107. https://doi.org/10.1038/s41467-020-16941-y.

Wilson, David C., Adebisi O. Araba, Kaine Chinwah, and Christopher R. Cheeseman. 2009. "Building Recycling Rates through the Informal Sector." *Waste Management* 29 (2): 629–635. https://doi.org/10.1016/j.wasman.2008.06.016.

Wilson, Japhy, and E. Swyngedouw, eds. 2014. *The Post-Political and Its Discontents: Spaces of Depoliticisation, Spectres of Radical Politics.* Edinburgh: Edinburgh University Press.

Wired. 2022. "A Circular Economy Means Track and Trace Transparency." February 3, 2022. https://www.wired.co.uk/bc/article/circular-economy-track-trace-transparency.

Wittmer, Josie. 2022. "Dirty Work in the Clean City: An Embodied Urban Political Ecology of Women Informal Recyclers' Work in the 'Clean City.'" *Environment and Planning E: Nature and Space*, May. https://doi.org/10.1177/25148486221102374.

Wittmer, Josie, and Kate Parizeau. 2016. "Informal Recyclers' Geographies of Surviving Neoliberal Urbanism in Vancouver, BC." *Applied Geography* 66 (January): 92–99. https://doi.org/10.1016/j.apgeog.2015.10.006.

Wittmer, Josie, Sharada Srinivasan, and Mubina Qureshi, 2020. "Women Waste Pickers' Lives during the COVID-19 Lockdown in Ahmedabad, India." SSRN. http://dx.doi.org/10.2139/ssrn.3885161.

Yates, Luke. 2015. "Everyday Politics, Social Practices and Movement Networks: Daily Life in Barcelona's Social Centres." *British Journal of Sociology* 66 (2): 236–258. https://doi.org/10.1111/1468-4446.12101.

Yates, Michelle. 2011. "The Human-as-Waste, the Labor Theory of Value and Disposability in Contemporary Capitalism." *Antipode* 43 (5): 1679–1695. https://doi.org/10.1111/j.1467-8330.2011.00900.x.

Yee, Vivian, and Hwaida Saad. 2019. "To Make Sense of Lebanon's Protests, Follow the Garbage." *New York Times*, December 3, 2019, sec. World. https://www.nytimes.com/2019/12/03/world/middleeast/lebanon-protests-corruption.html.

Zapata Campos, María José, Sebastián Carenzo, Jaan-Henrik Kain, Michael Oloko, Jessica Pérez Reynosa, and Patrik Zapata. 2021. "Inclusive Recycling Movements: A Green Deep Democracy from Below." *Environment and Urbanization* 33 (2): 579–598. https://doi.org/10.1177/0956247820967621.

Zapata Campos, María José, and Patrik Zapata. 2017. "Infiltrating Citizen-Driven Initiatives for Sustainability." *Environmental Politics* 26 (6): 1055–1078. https://doi.org/10.1080/09644016.2017.1352592.

Zimmer, Anna, Natasha Cornea, and René Véron. 2017. "Of Parks and Politics: The Production of Socio-Nature in a Gujarati Town." *Local Environment* 22 (1): 49–66. https://doi.org/10.1080/13549839.2016.1157157.

Zimring, Carl A. 2017. *Clean and White: A History of Environmental Racism in the United States*. New York: New York University Press.

Index

Page numbers followed by *t* indicate tables and *f* indicate figures.

Urban and Industrial Environments

Series editor: Robert Gottlieb, Henry R. Luce Professor of Urban and Environmental Policy, Occidental College

Jason Corburn, *Street Science: Community Knowledge and Environmental Health Justice*

Peggy F. Barlett, ed., *Urban Place: Reconnecting with the Natural World*

David Naguib Pellow and Robert J. Brulle, eds., *Power, Justice, and the Environment: A Critical Appraisal of the Environmental Justice Movement*

Eran Ben-Joseph, *The Code of the City: Standards and the Hidden Language of Place Making*

Nancy J. Myers and Carolyn Raffensperger, eds., *Precautionary Tools for Reshaping Environmental Policy*

Kelly Sims Gallagher, *China Shifts Gears: Automakers, Oil, Pollution, and Development*

Kerry H. Whiteside, *Precautionary Politics: Principle and Practice in Confronting Environmental Risk*

Ronald Sandler and Phaedra C. Pezzullo, eds., *Environmental Justice and Environmentalism: The Social Justice Challenge to the Environmental Movement*

Julie Sze, *Noxious New York: The Racial Politics of Urban Health and Environmental Justice*

Robert D. Bullard, ed., *Growing Smarter: Achieving Livable Communities, Environmental Justice, and Regional Equity*

Ann Rappaport and Sarah Hammond Creighton, *Degrees That Matter: Climate Change and the University*

Michael Egan, *Barry Commoner and the Science of Survival: The Remaking of American Environmentalism*

David J. Hess, *Alternative Pathways in Science and Industry: Activism, Innovation, and the Environment in an Era of Globalization*

Peter F. Cannavò, *The Working Landscape: Founding, Preservation, and the Politics of Place*

Paul Stanton Kibel, ed., *Rivertown: Rethinking Urban Rivers*

Kevin P. Gallagher and Lyuba Zarsky, *The Enclave Economy: Foreign Investment and Sustainable Development in Mexico's Silicon Valley*

David N. Pellow, *Resisting Global Toxics: Transnational Movements for Environmental Justice*

Robert Gottlieb, *Reinventing Los Angeles: Nature and Community in the Global City*

David V. Carruthers, ed., *Environmental Justice in Latin America: Problems, Promise, and Practice*

Tom Angotti, *New York for Sale: Community Planning Confronts Global Real Estate*

Paloma Pavel, ed., *Breakthrough Communities: Sustainability and Justice in the Next American Metropolis*

Anastasia Loukaitou-Sideris and Renia Ehrenfeucht, *Sidewalks: Conflict and Negotiation over Public Space*

David J. Hess, *Localist Movements in a Global Economy: Sustainability, Justice, and Urban Development in the United States*

Julian Agyeman and Yelena Ogneva-Himmelberger, eds., *Environmental Justice and Sustainability in the Former Soviet Union*

Jason Corburn, *Toward the Healthy City: People, Places, and the Politics of Urban Planning*

JoAnn Carmin and Julian Agyeman, eds., *Environmental Inequalities Beyond Borders: Local Perspectives on Global Injustices*

Louise Mozingo, *Pastoral Capitalism: A History of Suburban Corporate Landscapes*